ENVIRONMENTAL SCIENCE, ENGINEE

SEQUENCING BATCH REACTORS

AN OVERVIEW

ENVIRONMENTAL SCIENCE, ENGINEERING AND TECHNOLOGY

Additional books and e-books in this series can be found
on Nova's website under the Series tab.

WASTE AND WASTE MANAGEMENT

Additional books and e-books in this series can be found
on Nova's website under the Series tab.

ENVIRONMENTAL SCIENCE, ENGINEERING AND TECHNOLOGY

SEQUENCING BATCH REACTORS

AN OVERVIEW

LOIS K. MELLO
EDITOR

nova
science publishers
New York

NOTICE TO THE READER

Library of Congress Cataloging-in-Publication Data

ISBN: 978-1-53615-462-7

Published by Nova Science Publishers, Inc. † New York

CONTENTS

PREFACE

Sequencing Batch Reactors: An Overview opens with the results of an investigation with the goal of determining the most suitable treatment for tannery effluents. The investigation was carried out on three distinct effluents produced by a tannery located in Venezuela, as well as a mixture of the three.

Next, the authors treated leachate originated from the Ouled Berjal landfill by using the sequential batch reactor technique. The leachate was put into two reactors, which differed in the times allocated to each phase.

The closing study presents an overview of studies using anaerobic sequencing batch biofilm reactors digesting agroindustry wastes for methane production, focusing on operational strategy and perspectives for scale-up.

Chapter 1 - The tannery industry is responsible for the transformation of animal skins into a stable and imputrecible material known as leather. During the production process, large quantities of different chemical reagents, mixed with water, are added; as a consequence, the associated effluents, in the form of wastewater, contain many organic and inorganic contaminants. These effluents need to be treated properly to remove these contaminants, thus ensuring a low environmental impact. This chapter shows the results of an investigation to determine the most suitable treatment for tannery effluents. The investigation was carried out on effluents produced by a tannery located in Venezuela. Three distinct effluents, as well as a mixture of all three (specifically, soak liquor, tanning, dyeing, and a mixture of these) were studied. Removal of organic matter and nutrients was evaluated using the combination of biological treatment in an SBR and physicochemical treatment. The three effluents, when mixed together, showed the best tratability characteristics; the mixture contained 57.4% organic matter, which is susceptible to removal by biological treatment. COD fractionation showed that the treatment sequence should combine a biological process in a SBR with a physicochemical treatment (coagulation-flocculation). To achieve the objectives, four different aeration sequences were tested during the experimentation, combined with three different SBR filling times (rapid, slow and step filling process). SBR operating conditions were HRT of 30 h, cycles of 12 h and sludge age of 15 d. The most efficient aeration sequence was predenitrification, combined with rapid filling of the reactor (5 min). A jar test was used to perform the physicochemical treatment. The optimal dose of coagulant was 60 mL·L^{-1} FeCl$_3$ (1.62 g·L^{-1}) and 5 mL·L^{-1} of bitter brine (100 g·L^{-1}) as an adjuvant. The treatment in the SBR was able to remove 52.8 to 53.4% of tCOD, representing an efficiency of 92 to 96.3% of the biological system and 60.6% for TN, while the remaining inert colloid material was removed by post-treatment. tCOD and NT concentrations met the limits set by Venezuelan law (303.4 and 35.1 mg·L^{-1}, respectively). According to the results, the proposed pilot-scale treatment included a primary treatment for homogenization, sedimentation and removal of chromium from the effluent, followed by biological treatment in a SBR and finally a post-treatment based on coagulation-flocculation.

Chapter 2 - Solid waste management is one of the major issues faced by countries around the world. The waste management approach used in Morocco consists mainly to improve the conditions for waste collection, sorting, and treatment while limiting the nuisance caused by methane emissions and leachate production.

In the present work, the authors have treated leachate originated from Ouled Berjal landfill (Kenitra -Morocco) by using the Sequential Batch

Reactor (SBR) technique. Thus, the leachate was put into two reactors, which differed in the times allocated to each phase. The authors have assessed the treatment effectiveness by monitoring the removal of NH_4^+ ions, nitrates, nitrites, phenol, absorbance at 254 nm and phytotoxicity. The results showed that the first SBR is suited for leachate treatment with removal efficiencies reaching 98,5% for NH_4^+ ions, 53% for nitrates, 61,5% for nitrites, 82,5% for phenol and 66% for the absorbance. In addition, the phytotoxicity tests showed an increase in the germination index from 19% to 47%, indicating that significant part of the pollutants was removed from the leachate.

Chapter 3 - Anaerobic digestion allows organic matter conversion into final products such as methane by microorganism activity. Over the last decades, engineering has adapted this process into anaerobic bioreactors with different configurations for wastewater treatment aimed at biogas production. Among reactor types, the anaerobic sequencing batch reactor (ASBR) is one of the several high rate configurations and it appears as an alternative to continuous systems. It presents advantages such as better effluent control and simple operation comprised in four steps: feed, reaction, settling of granular biomass and decant. Another configuration consists in the anaerobic sequencing batch biofilm reactor (AnSBBR), in which the biomass is immobilized in inert support. It enables the elimination of the settling step, thus reducing the overall cycle time. Although they have different biomass arrangements, the operational factors tend to be the same: agitation, food/microorganism ratio, reactor configuration and feed strategy. AnSBBRs have applied to mesophilic and thermophilic treatment of wastewaters from agroindustry such as vinasse (bioethanol production), whey (dairy industry), and glycerin (biodiesel production) with various operational strategies: feeding mode, temperature, organic load, influent concentration and cycle time. Therefore, this study presents an overview of achievements of studies that used AnSBBRs digesting agroindustry wastes for methane production, focused on operational strategy and perspectives for scale-up estimative.

In: Sequencing Batch Reactors: An Overview   ISBN: 978-1-53615-462-7
Editor: Lois K. Mello   © 2019 Nova Science Publishers, Inc.

Chapter 1

TANNERY WASTEWATER TREATMENT APPLYING BIOLOGICAL AND PHYSICOCHEMICAL PROCESSES

María Carolina Pire-Sierra[1,], PhD*
and Sedolfo José Carrasquero-Ferrer[2], PhD
[1]Ecology and Quality Control Department,
Centroccidental University, Barquisimeto, Lara, Venezuela
[2]Sanitary and Environmental Engineering Department,
La Universidad del Zulia, Maracaibo, Zulia, Venezuela

ABSTRACT

The tannery industry is responsible for the transformation of animal skins into a stable and imputrecible material known as leather. During the production process, large quantities of different chemical reagents, mixed with water, are added; as a consequence, the associated effluents, in the form of wastewater, contain many organic and inorganic contaminants. These effluents need to be treated properly to remove these contaminants, thus ensuring a low environmental impact. This chapter shows the results of

* Corresponding Author's E-mail: mcpirre@ucla.edu.ve

an investigation to determine the most suitable treatment for tannery effluents. The investigation was carried out on effluents produced by a tannery located in Venezuela. Three distinct effluents, as well as a mixture of all three (specifically, soak liquor, tanning, dyeing, and a mixture of these) were studied. Removal of organic matter and nutrients was evaluated using the combination of biological treatment in an SBR and physicochemical treatment. The three effluents, when mixed together, showed the best treatability characteristics; the mixture contained 57.4% organic matter, which is susceptible to removal by biological treatment. COD fractionation showed that the treatment sequence should combine a biological process in a SBR with a physicochemical treatment (coagulation-flocculation). To achieve the objectives, four different aeration sequences were tested during the experimentation, combined with three different SBR filling times (rapid, slow and step filling process). SBR operating conditions were HRT of 30 h, cycles of 12 h and sludge age of 15 d. The most efficient aeration sequence was predenitrification, combined with rapid filling of the reactor (5 min). A jar test was used to perform the physicochemical treatment. The optimal dose of coagulant was 60 mL·L^{-1} FeCl$_3$ (1.62 g·L^{-1}) and 5 mL·L^{-1} of bitter brine (100 g·L^{-1}) as an adjuvant. The treatment in the SBR was able to remove 52.8 to 53.4% of tCOD, representing an efficiency of 92 to 96.3% of the biological system and 60.6% for TN, while the remaining inert colloid material was removed by post-treatment. tCOD and NT concentrations met the limits set by Venezuelan law (303.4 and 35.1 mg·L^{-1}, respectively). According to the results, the proposed pilot-scale treatment included a primary treatment for homogenization, sedimentation and removal of chromium from the effluent, followed by biological treatment in a SBR and finally a post-treatment based on coagulation-flocculation.

Keywords: sequential batch reactor (SBR), biological nutrient removal, physicochemical treatment.

CHARACTERIZATION OF THE EFFLUENTS PRODUCED IN THE TANNERY

Characteristics of wastewater produced in tanneries vary depending on the company size, the types of chemical reagents used, the amount of water and the type of final product (Durai and Rajasimman, 2011). The studied tannery generated effluents in batches, producing individual fractions of wastewater (soak liquor, tanning and dyeing) which were stored in ponds. In this study, physicochemical characterization of all effluents produced in the tannery is presented.

Several researchers agree that these effluents must be segregated when they are generated, in order to recover specific compounds of each stream, such as sulfides, characteristic of the process of soak liquor, and chromium, a typical by-product of the tanning process (Kabdasli et al., 1993; Orhon et al., 1998; Lefevbre et al., 2005).

Soak Liquor Fraction (P)

The results of the characterization of soak liquor, the first fraction of the effluents generated in the leather manufacturing process, are presented in Table 1. It is observed that the effluent has a basic character (pH=12) with light brown coloration and high organic matter concentration (64,253 - 81,617 mg $COD·L^{-1}$), a high content of nitrogenous compounds (5,488 - 6,500 mg $TKN·L^{-1}$) and total suspended solids (11,810 - 23.200 mg $TSS· L^{-1}$).

The high concentration of contaminants in the soak liquor fraction comes from the cleaning process of skins, used in order to eliminate traces of blood, meat (proteins) and fat (lipids). The high VSS/TSS ratio (0.9 $mg·mg^{-1}$) corresponds to the content of organic matter present in the effluent. The basic pH was the result of the addition of lime, sodium sulfide and ammonia, required to remove hair from the skins (Nemeron, 1977; Lefebvre et al., 2006).

The concentrations of total COD (tCOD) and total Kjeldahl nitrogen (TKN) were high due mainly to the high amount of organic matter and nitrogen in the soak liquor fraction. This fact was reported by Kabdasli et al. (1993) when they worked with segregated effluents associated with the processing of fur and the supernatant of skin washes, attributing this behavior to the presence of a significant content of recalcitrant particulate matter.

The COD/TKN ratio was 12.2 $mg·mg^{-1}$, which means the effluent had a carbon to nitrogen ratio adequate to permit biological treatment, according to

Carucci et al. (1999) and Szpyrkowicz and Kaul, (2004) who used effluents from tanneries and found that this ratio should be higher than 8 and 7 mg·mg[-1], respectively. However, the low content of ammoniacal nitrogen indicates that most of the nitrogen present in the effluents comes from the proteins and fats of the skins, which are complex organic compounds that are slow to be ammonified (Ekama and Wentzel, 2008a; Insel et al., 2009). This is an important aspect to consider if a biological wastewater treatment will be used.

Table 1. Conventional characterization of the effluents produced in the tannery ($\overline{X} \pm SD$)

Parameter	Soak liquor (P)	Tanning (C)	Dyeing (T)	Mix of fractions stored in a pond (L)
pH	12.0 ± 0.3	3.5 ± 0.1	3.3 ± 0.7	8.2 ± 0.5
Alcalinity (mgL[-1])	11,233 ± 94.7	-----	-----	13,400 ± 2,900
Acidity (mg·L[-1])	----	5,532.4 ± 145.4	2,539.3 ± 1,029.2	-----
TSS (mg·L[-1])	17,505 ± 5,695	10,095 ± 2,457	2,427 ± 1,072	2,283 ± 308.1
VSS (mg·L[-1])	15,150 ± 6,820	3,240 ± 994.4	787 ± 270.5	886 ± 181.2
VSS/TSS	0.9 ± 0.1	0.3 ± 0.1	0.3 ± 0.1	0.4 ± 0.2
tCOD (mg·L[-1])	72,935 ± 8,681.9	8,843 ± 2,931.0	7,680 ± 4,590.6	2,510 ±423.7
sCOD (mg·L[-1])	----	----	----	1,594.9 ± 120.7
$BOD_{5,20}$ (mg·L[-1])	----	----	----	1,053.1 ± 134.1
BOD/tCOD	----	----	----	0,42
TKN (mg·L[-1])	5,994 ± 505.9	361.8 ± 248.7	188.2 ± 10.5	171.2 ± 91.0
$N-NH_4^+$ (mg·L[-1])	660.3 ± 323.4	271.4 ± 126.7	56.0 ± 32.5	127.1 ± 137.9
$N-NO_2^-$ (mg·L[-1])	ND[1,2]	ND[1,2]	ND[1,2]	1.3 ± 0.9
$N-NO_3^-$ (mg·L[-1])	ND[1,2]	ND[1,2]	ND[1,2]	2.7 ± 3.0
$P-PO_4^{3-}$ (mg·L[-1])	151.3 ± 28.2	17.8 ± 7.7	26.8 ± 9.6	----
Total chromiun (mg·L[-1])	0 ± 0.05	492.4 ± 62.9	58.2 ± 39.2	6.3 ± 0,5

Pire-Sierra et al. (2011); Note: [1]ND: not detectable. [2] Detection limit: 1 mgNO$_x$·L[-1].

The experimental results obtained for the soak liquor fraction were similar to those obtained in previous investigations by Pire-Sierra et al. (2010a), when this effluent was first characterized; however, results shown in Table 1 were markedly higher than those found by Lefebvre et al. (2005), who used the supernatant of the soak liquor, and reported a COD concentration of 2,200 mg·L^{-1}, a TSS of 5,300 mg·L^{-1}, a TKN of 273 mg·L^{-1} and a pH of 7.7. Despite the difference in the results obtained, it was observed that in both investigations, the COD/TKN ratio was above 8 mg·mg^{-1} and the COD/PO$_4^{3-}$ ratio was lower 1 mg·mg^{-1}, which is characteristic of tannery effluents.

Tanning Fraction (C)

The second fraction of wastewater produced in the tannery is called *tanning*, which is characterized by its acidity (pH=3.5), an intense blue color, due to the presence of trivalent chromium dissolved in the aqueous medium. Acid media is due to the addition of sulfuric acid during the production process, which is required to keep trivalent chromium soluble and avoid its precipitation on skin fibers (Nemeron, 1977; Farabegoli et al., 2004). The content of organic matter was 8,843 mg COD·L^{-1} and of TKN of 361 mg·L^{-1} (Table 1).

The levels of trivalent chromium remaining in the effluent ($\approx$492.4 mg·L^{-1}) indicate a waste of this metal during the tanning process. Kabdasli et al. (1993) pointed out that chrome discharges higher than 460 mg·L^{-1} indicating an inefficiency during the tanneries productive process, which generates losses of this heavy metal. To prevent this, they suggest making adjustments in the process to allow for greater adsorption of the metal in the skin, as well as for reduction in the dose of chromium salts used, concomitantly decreasing the excessive discharge of chromium in the effluent.

The VSS/TSS ratio was 0.3 mg·mg^{-1}, which indicates a high content of inorganic material in the tanning effluent, typical of the diversity of chemical reagents that are added during this stage of the production process of leather. On the other hand, the COD/TKN ratio was 24.4, which is considered adequate for the biological elimination of nutrients (Abu-Gharlarah and Randall, 1990; Carucci et al., 1999; Szpyrkowicz and Kaul, 2004), however, due to the high content of inorganic material in the effluent, it is considered a low biodegradability wastewater (Lefebvre et al., 2005; Ganesh et al., 2006; Durai and Rajasimman, 2011).

An important part of the inorganic material in the tanning fractions is a high concentration of chromium, a component classified as toxic and inhibitory to microbial processes (Insel et al., 2006; Karahan et al., 2008). In this regard, Farabegoli et al. (2004) found that concentrations higher than 180 mg $Cr \cdot L^{-1}$ produce almost total inhibition of nitrifying microorganisms, preventing the biological removal of nitrogen from the effluents. To treat wastewater from the tanning process, the chromium content in this fraction would have to be reduced using chemical sedimentation. Thus, the pH is directly related to the removal of chromium from the effluent and its subsequent recovery, which is necessary in order to reduce the production of the toxic sludge generated from the process (Kabdasli et al., 1993; Farabegoli et al., 2004; Kanagaraj et al., 2008).

Dyeing Fraction (T)

The third fraction of the effluent produced by the tannery is generated of the finishing process, called *dyeing*. The coloration of the effluent depends on the dyeing color of the leathers, varying between black, dark brown and dark blue. This fraction has an acid character (pH of 3.3) with average concentrations of COD, TKN and $P-PO_4^{3-}$ of 7,680 $mg \cdot L^{-1}$, 188.2 $mg \cdot L^{-1}$ and 30.5 $mg \cdot L^{-1}$, respectively (Table 1). The VSS/TSS ratio was 0.3 $mg \cdot mg^{-1}$, indicating that it is an effluent with a high content of inorganic material, coming from the coloring agents and dyes used during the final stage of the process.

The low pH value and the high acidity observed in this fraction is due to the fact that during the dyeing stage, the skin is discolored by adding sulfuric acid to clarify and uniformize its color, thus making the application of the dye more efficient (Nemerow, 1977). The organic matter and nitrogen contained in these effluents come mainly from the products used in this stage. For example, AZO dyes and metal complexes, oils, dispersants, retanning agents, fats, waxes, pigments and alcohols, used during the filling stage, so that the leather is more foldable and resistant to tearing. These chemical products in the wastewater are contaminants, with a high percentage of components resistant to biodegradation (Galisteo et al. 2002; Stoop, 2003).

In previous investigations, Pire-Sierra et al. (2010a, 2010b) characterized the effluent of the dyeing fraction from the tannery and obtained similar values for the COD and pH parameters (8,891.1 mg·L^{-1} and 3.2, respectively). However, they obtained different results for TKN (542 mg·L^{-1}) and for acidity (2,425 mg·L^{-1}). These differences are due to the variation of the production process, and the different chemicals present in the AZO dyes and pigments used to dye the skins, which are mainly responsible for the presence of nitrogen compounds in the effluent (Kabdasli et al., 1993; Galisteo et al., 2002).

The COD/NTK ratio for the dyeing fraction was 40.8, representing an adequate ratio between organic matter and nitrogen (Abu-Ghararah and Randall, 1990; Caruccí et al., 1999; Szpyrkowicz and Kaul, 2004). The characterization of the dyeing, the last fraction of the effluent produced in the tannery, showed that it was the less toxic effluent of the three studied. It did not contain sulfides, which are characteristic of the effluent of the fur, and at the same time, the concentration of chromium was significantly lower than that reported for the tanning fraction (58.2 mg Cr·L^{-1}). Thus, the only significant pollutants for the dyeing effluent were COD and total nitrogen.

The presence of NO_2^- and NO_3^- was not detected for any of the individually characterized fractions, due to the fact that the residual water was recently produced. This wastewater had high oxygen demand, as well as presence of the toxic compounds at the concentrations reported in Table 1. These conditions make it difficult to treat tannery wastewater using biological processes (Barajas, 2002; Durai and Rajasimman, 2011).

The characteristics of the tannery effluents indicate that they require a pre-treatment before applying biological treatment. Hence, the soak liquor fraction should be subjected, for example, to a catalytic oxidation process with manganese sulfate; to remove the sulfides present (Munz et al., 2008). On the other hand, for the tanning fraction, the chromium content would have to be reduced and subsequently recovered, in order to reuse it in the process, and in order to reduce the production of toxic sludge (Kabdasli et al. 1993; Farabegoli et al. 2004; Kanagaraj et al., 2008). Finally, pH of all the fractions was outside the acceptable range for a biological systems (6 - 9 units), so it has to be adjusted using a of homogenization unit prior to the biological treatment (Metcalf and Eddy, 1995).

All of the treatments mentioned above would increase operating costs, but would not guarantee good removal efficiencies in the biological system, due to the apparent low biodegradability presented by the individually characterized fractions. Based on this fact, it was considered important to study the characteristics of these mixed fractions in the proportions in which they are actually produced in the tannery.

Mix of Fractions Stored in a Pond (L)

As mentioned, the three individual effluents produced in the tanning process, together with the water used for cleaning in the industry, are stored in a pond. It was found that the mixture is slightly basic, with a pH of 8.2. The coloration of the effluent was reddish, produced from the mixture of dyes and colorants used during the finishing process of the leathers. The conventional characterization showed that the mixture stored in the pond had a COD concentration of 2,510.2 mg·L^{-1}, for the TKN of 171.2 mg·L^{-1} and for trivalent chromium of 6.3 mg·L^{-1} (Table 1), representing the lowest concentrations of pollutants, among the individual fractions.

The results shown in Table 1 indicate that storing wastewater in a pond worked as a pretreatment because homogenization and sedimentation processes takes place, and results in remotion of contaminants (Pire-Sierra et al., 2011). Wastewater pH stabilizated at 8 units during storage in the pond; under this condition chromium is not soluble (Farabegoli et al., 2004) and therefore, it precipitates along with other sedimentable components, causing a decrease of these pollutans from the liquid fraction of the wastewater (Orhon et al., 1999a).

The content of ammoniacal nitrogen was 127.1 mg·L^{-1}, which represents 74.2% of TKN, meaning that organic nitrogen was around 32.8 mg·L^{-1}

(25.8% of TKN). These results show that most of the nitrogen presents in the effluent storaged in the pond was already ammonified. On the other hand, the concentration of total suspended solids (TSS) was 2,310 $mg \cdot L^{-1}$ and the VSS/TSS ratio was 0.5 $mg \cdot mg^{-1}$, indicating a wastewater with a moderate organic load. The COD:TKN:P-PO_4^{3-} relation obtained was 100:9.7:0.5, which is within recommendations made by other researchers, as minimum requirement to apply biological treatment (Carucci et al., 1999; Szpyrkowicz and Kaul, 2004); however, the biodegradability index ($BOD_{5,20}/tCOD$) was 0.42, and it was classified as an effluent with low biodegradability, according to the classification of Ahn et al. (1999) and INESCOP (2008).

Studies conducted by Graterol (2000), using an effluent from a tannery located in the central-western region of Venezuela, reported concentrations of 2,500 mg $COD \cdot L^{-1}$, 2,500 mg $TSS \cdot L^{-1}$ and pH of 9 units, values very similar to those obtained in the present investigation, but the concentration of TKN was slightly lower (120 $mg \cdot L^{-1}$). However, Munz et al. (2008), who characterized the effluent of a tannery taken from a primary sedimentation process with oxidized sulphides, found concentrations of COD and TKN (3,690 and 306 $mg \cdot L^{-1}$, respectively) higher than those found in the present investigation (Table 1), while the values of pH and TSS (6.9 and 976 $mg \cdot L^{-1}$, respectively) were lower. The variations in the characteristics of the effluents are due to changes in the chemical compounds used in the production processes, and the type of leather produced (Kabdasli et al., 1993; Galisteo et al., 2002; Durai and Rajasimman, 2011).

The content of suspended solids was high in all the effluents studied (P, C, T and L), coinciding with that indicated by Ganesh et al. (2006), Lefebvre et al. (2006) and Durai and Rajasimman (2011) who affirmed that it is a intrinsic characteristic of tannery effluents, which can oscillate between 5,300 and 10,700 $mg \cdot L^{-1}$. It is also important to note that the individual effluents did not show the presence of nitrites or nitrates (limit of detection of 1 $mg \cdot L^{-1}$), while the effluent stored in the pond showed low concentrations of these ions, thus indicating a possible partial nitrification process.

All the effluents of the tannery studied showed a deficiency of total phosphorus, revealed by the values of the relation of COD:TKN:P-PO_4^{3-}. This deficiency is a typical characteristic of tannery effluents, but in most cases, it was not a problem to carry out the biological treatment of the effluents (Karahan et al., 2008; Durai and Rajasimman, 2011), only in some cases, it was necessary to add phosphorus as a complement to the residual water for application of biological treatment (Ryu et al., 2007; Munz et al., 2008).

The mixing of the fractions of the effluents in the storage pond improved some of their characteristics, in terms of concentration of contaminants and the relationships between their nutrients (Pire-Sierra et al., 2011). However, according to the classification presented by Ahn et al. (1999) and INESCOP (2008), when evaluating the biodegradability relation, it was found that the L mixture is not very biodegradable (BOD/COD = 0.42). Based on this result, the application of a biological treatment system in wastewater from tanneries would not be viable. However, researchers such as Vidal et al. (2004) and Lefebvre et al. (2005) successfully applied biological treatments in secuential batch reactors (SBR) using effluents from tanneries with BOD/COD ratios of 0.3, which indicated that this way to measure biodegradability was not completely adequate to know the biodegradability in this effluent.

Conventional characterization was not enough to define the effluent with the best biodegradability. Several researchers agree that in order to establish the biodegradability of effluents, particularly industrial effluents, a deeper study of COD, known as COD fractionation, should be done (Park et al., 1997; Orhon et al., 1999a; Ekama and Wentzel, 2008b; Karahan et al., 2008). By means of this test it is possible to determine the content of biodegradable and non-biodegradable material of the effluents.

Fractionation of Chemical Oxygen Demand

The fractionation of the COD was a useful tool in the determination of the biodegradability of the wastewater, as it permitted measurement of the content of biodegradable and non-biodegradable organic matter. Likewise, the subclassification of the COD components of the effluents was obtained, according to their speed of degradation and their soluble or particulate contents (Park et al., 1997; Orhon et al, 1999a; Ekama and Wentzel, 2008b).

Based on the information generated by the fractionation of the tannery effluents, it was possible to define the most convenient wastewater treatment for the effluents stored in the pond (Galisteo et al., 2002; Karahan et al., 2008).

The conventional biodegradabily ratio, $BOD_{5,20}$/tCOD, is 42% (Table 1); but it was found to be less than 57.4% of the biodegradable material (CODBT) obtained by the COD fractionation test (Table 2). These results showed that the total COD concentration in the wastewater did not provide information on the effluent's biological treatability (Yildiz et al., 2008). Similar results were obtained by Arslan and Ayberk (2003), who found that the biodegradability ratio for a mixture of industrial and domestic effluents ranged between 0.09 and 0.36, so that the effluent would be classified as a non-biodegradable mixture. However, when they measured the fractionation of COD of this mix, it was found that a large part of the organic content in the effluent was biodegradable, oscillating between 84 and 92% of the total COD. They concluded that the conventional characterization was not reliable enough to design a biological treatment for these effluents.

The characterization and fractionation of the COD of the wastewater stored in the pond showed that the tCOD was 2,510.2 mg·L^{-1} (Table 1), with a biodegradable fraction of 57.4% and a non-biodegradable fraction of 42.6% (Table 2). These results show the difference between the tCOD and the CODBT; hence the importance of knowing the COD fractions, in order to be able to make a more appropriate design of the biological treatment systems, since the design would be based on the content of the organic biodegradable materials, instead of total organic matter (Cokgor et al., 1998; Boursier et al., 2005; Palmero et al., 2009).

The results obtained from the fractionation of the COD were found to be within the range indicated in the literature for these effluents. In this regard, Insel et al. (2009) found that the biodegradable fraction of sedimented effluent from a tannery was 81%, while Karahan et al. (2008) found that only 34.9% of the effluent was biodegradable. These results show the variability in the biodegradability presented by the effluents, attributed to the difference of chemical compounds added during each process, and the type of processed skins (Galisteo et al., 2002; Durai and Rajasimman, 2011), as well as to the existence or not of a pre-treatment, such as sedimentation (Orhon et al., 1999).

Table 2. Fractionation of the COD for the effluents produced in the tannery*

COD fractionation	Symbol	Mix of fractions stored in a pond (L)
General (% ± SD)		
Total biodegradable COD	CODBT	57.4 ± 10.0
Total no biodegadable COD	CODNBT	42.6 ± 10.0
Detailed (%)		
Easily biodegradable COD	CODFB	33.1 a
Slowly biodegradable COD	CODLB	24.3 a
Soluble no biodegradable COD	CODNBs	29.0 b
Particulate no biodegradable COD	CODNBp	13.6 c

*Pire-Sierra et al., 2011.
Mean followed by different letters in each row indicates significant differences according to the Tukey test (P≤0.05).

The easily biodegradable COD (CODFB) for the effluent L (stored in the pond) was 33.1%, higher tan the 24.3% of the slowly biodegradable COD (CODLB) obtained for the same effluent. The CODFB was higher than that reported by other researchers, who also used wastewater from tanneries for which the range was between 7.5 and 19% (Orhon et al., 1999a; Karahan et al., 2008; Insel et al., 2009). However, it behaved similarly to the CODFB reported by Palmero et al. (2009), who performed the fractionation of the COD to a poultry effluent. They found that the CODFB (58%) was superior to the CODLB (31%) for a total COD of 1,840.4 mg·L^{-1}.

The importance of knowing the content of CODFB in industrial wastewater is that this type of organic matter is required by microorganisms for the processes of biological nitrogen removal (Von Sperling and Lemos, 2006). During the denitrification stage, approximately 7 g of CODFB are required to remove 1 g of NO_3^- (Insel, 2007). Therefore, it is desirabale to use the effluent that has the highest CODBT content to carry out the biological treatment, particularly the one with the highest easily biodegradable fraction, because the COD is immediately available for microorganisms, while the slowly biodegradable fraction must be first hydrolyzed by microorganisms, which requires more time (Park et al., 1997; Boursier et al., 2005; Ekama and Wentzel, 2008b).

COD fractionation showed that the slowly biodegradable fraction of the COD was 24.3%, similar to that obtained by Karahan et al. (2008) for effluents from tanneries, who reported 27.6% of CODLB. However, other researchers found that the slowly biodegradable fraction in tannery effluents was around 60% of the tCOD (Cokgor et al., 1998; Insel et al., 2009). The importance of knowing the CODLB is related to the duration of the biological treatment (Orhon et al., 1999c); shorter treatment durations were required for effluents with low CODLB content, because the fraction that must be previously hydrolyzed by the microorganisms is less.

The total non-biodegradable COD content (CODNBT) in the effluent stored in the pond was 42.6%, consisting mostly of the inert soluble COD (29%), while the remaining 13.6% corresponded to the inert particulate fraction. The content of inert soluble material is responsible for the low removal efficiencies reported by some researchers when they applied biological treatment to tannery effluents (Pire-Sierra et al., 2010a, 2010b), because this fraction of the COD cannot be removed by the biological system. Only physicochemical treatments, such as adsorption, coagulation-flocculation and physical entrapment are able to remove the colloidal components of this fraction (Dosta et al., 2008; Karahan et al., 2008).

On the other hand, the importance of knowing the CODNBp lies in the fact that it is the fraction capable of intertwining with the biomass of a biological system, and separating from wastewater by sedimentation (Ekama and Wentzel, 2008b). However, it would only be removed from the system by controlling the age of the sludge (purge). This would imply carrying out adequate purge volumes to avoid the accumulation of this particulate COD in the biological system. However, the selection of the sludge ages would not depend only on the content of the CODNBp, but also on the objective of the biological treatment (Palma and Manga, 2005; Ekama and Wentzel, 2008b; González and Sadarriaga, 2008).

The CODNBp content of the effluent storaged in the pond was 13.6%, similar to that reported by Orhon et al. (1999a) and Insel et al. (2009) who obtained 11.0 and 11.5% of this fraction, likewise it was within the wide range of variation of the CODNBp reported by Karahan et al. (2008) and Insel et al. (2009), which ranged from 11.0 to 52.2% of the total COD.

Hermida-Veret et al. (2000) noted that chemical sedimentation was able to remove part of this inert particulate COD, which improved the quality of the wastewater before being fed to a biological treatment system; however, the effluent still contained a high percentage of CODNBp. In this regard, Orhon et al. (1999a) obtained an effluent with almost zero particulate content, after applying a chemical sedimentation to the effluent.

The results of this investigation showed the difference between the content of organic matter measured as COD and as BOD, of the effluents stored in the pond (Figure 1). It was observed that the CODBT is lower than the tCOD, due to the presence of inert material. Likewise, $BOD_{5,20}$ was significantly lower than the content of biodegradable material (CODBT). The behavior of $BOD_{5,20}$ reinforced the theory that BOD is not a good parameter when tannery wastewater is used, due to the interference caused by the presence of toxic residues and inhibitors of tannery effluents (Metcalf and Eddy, 1995; Vidal et al., 2004; Lefebvre et al., 2005).

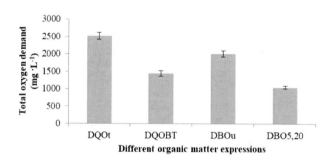

Figure 1. Different ways of expressing the content of organic matter for the effluent stored in the pond. Vertical bar represents the typical error.

Detailed Characterization of the Effluent with Better Biodegradability Characteristics

Characterization of the wastewater stored in the pond was carried out continuously over a period of two years (May, 2009 to May, 2011) where the profile of the effluent produced was determined in detail, during the rainy and dry season (Table 3). The tannery industry has two storage ponds whose dimensions are $\approx 1,400$ m^3 (pond 1) and $\approx 2,900$ m^3 (pond 2), they are outdoors. Intense rainy periods such as those registered during September 2010 and March 2011, influenced the concentration of contaminants present in the ponds, together with a decrease in the production of the tannery.

Table 3. Characterization of the wastewater stored in the pond during rainy season and dry season

Variable	Rainy season $\bar{X} \pm SD^1$ (C.V.)^2	Dry season $\bar{X} \pm SD^1$ (C.V.)^2
pH	8.1±0.4 (0.0)	8.2 ± 0.5 (0.1)
^{3}TSS (mg·L^{-1})	2004.8 ± 342.5 (0.2)	2282.7 ±308.1 (0.1)
^{4}VSS (mg·L^{-1})	756.2 ± 110.6 (0.1)	885.7 ± 181.8 (0.2)
^{5}tCOD (mg·L^{-1})	1394.5±204.0 (0.1)	2510.2 ± 418.5 (0.2)
^{6}sCOD (mg·L^{-1})	724.4 ± 120.7 (0.2)	1594.9 ± 452.4 (0.3)
^{7}TKN (mg·L^{-1})	137.7 ± 24.6 (0.2)	171.2 ± 91.0 (0.5)
^{8}N-NH$_4^+$ (mg·L^{-1})	77.0 ± 16.0 (0.2)	127.1 ± 137.9 (1.2)
Cr (mg·L^{-1})	3.8 ± 0.9 (0.2)	6.3 ± 0.5 (0.4)

Note: [1]S.D.:Standard deviation, [2]C.V: Coefficient of variation, [3]TSS: Total suspended solids, [4]VSS: Volatile suspended solids, [5]tCOD: Chemical total oxygen demand, [6]sCOD: Chemical soluble oxygen demand, [7]TKN: Total Kjeldahl nitrogen, [8]N-NH$_4^+$: Ammoniacal nitrogen.

The soluble COD (sCOD) of the effluent storaged in the pond represented between 50% and 60% of the total COD (tCOD). The regression equation and its coefficient of determination indicated that approximately 81% of the variation in COD directly affected the variation in COD (Figure 2). In parallel, the coefficients of variation for the tCOD and sCOD indicated in Table 3 showed that the variability of the concentrations measured during the two years of study was low.

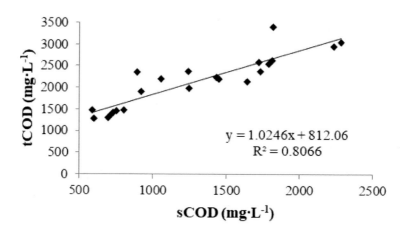

Figure 2. Association between total and soluble chemical oxygen demand (tCOD and sCOD) of the effluent stored in the pond (n ≥ 22 data).

Table 4. Pearson correlation matrix for the COD components of the effluent stored in the pond

	tCOD[1]	sCOD[2]	CODp[3]	CODBT[4]	CODNBT[5]	CODFB[6]	CODLB[7]	CODNBs[8]
sCOD	0.7926							
P*	0.0036							
CODp	0.4951	-0.0557						
P*	0.1215	0.8709						
CODBT	0.8660	0.9318	0.0601					
P*	0.0006	0.0000	0.8606					
CODNBT	-0.2928	-0.7013	0.5598	-0.7318				
P*	0.3822	0.0162	0.0733	0.0105				
CODFB	0.8192	0.8416	0.1151	0.9459	-0.6921			
P*	0.0020	0.0012	0.7361	0.0000	0.0183			
CODLB	0.7825	0.8947	-0.0226	0.9037	-0.6614	0.7159		
P*	0.0044	0.0002	0.9475	0.0001	0.0267	0.0132		
CODNBs	-0.3779	-0.4993	0.1285	-0.6077	0.6469	-0.6519	-0.4474	
P*	0.2519	0.1179	0.7066	0.0473	0.0315	0.0297	0.1676	
CODNBp[9]	-0.0213	0.4380	0.6072	-0.3744	0.6868	-0.2806	-0.4355	-0,1101
P*	0.9503	0.1778	0.0475	0.2566	0.0196	0.4033	0.1807	0.7474

[1]tCOD: Total chemical oxygen demand, [2]sCOD: Soluble chemical oxygen demand, [3]CODp: Particulate chemical oxygen demand, [4]CODBT: Total biodegradable chemical oxygen demand, [5]CODNBT: Total non-biodegradable chemical oxygen demand, [6]CODFB: Easily biodegradable chemical oxygen demand, [7]CODLB: Slowly biodegradable chemical oxygen demand, [8]CODNBs: soluble non-biodegradable chemical oxygen demand, [9]CODNBp: Particulate non-biodegradable chemical oxygen demand.

* Pearson's correlation is significant when P≤0.05.

To know in more detail the relationship of the COD of the effluent stored in the pond with each one of its COD fractions, the Pearson correlation matrix was made, in which the value of the Bonferroni probability was shown for each correlation (Barajas, 2002). First, significance was obtained between the CODBT and the tCOD with a positive correlation coefficient (r = 0.87), so that as the tCOD of the effluent increases, the CODBT also increases (Table 4).

For the CODBT, significant correlations of the biodegradable components (CODFB and CODLB) were obtained (p ≤ 0.05), indicating that the CODBT was more dependent on the CODFB, with a Pearson coefficient of 0.95, than of the CODLB, which presented a slightly lower coefficient (r = 0.90). Likewise, there was a significant correlation between sCOD and CODBT, with a correlation coefficient of 0.93. Of the components of the biodegradable COD, the CODLB had the greatest influence on the soluble COD (r = 0.89).

Regarding the non-biodegradable fraction of the COD, it was found that the sCOD and CODBT were negatively influenced by the CODNBT with a Pearson coefficient of -0.70, implying that an increase in the sCOD or CODBT would represent a decrease of the CODNBT. It was also possible to determine that the CODNBT was mainly influenced by the particulate fraction of the non-biodegradable COD, presenting a correlation coefficient of 0.69.

The Pearson correlation matrix indicated that the main component in the CODLB was the soluble part, since a significant relationship was obtained between sCOD and CODLB with a correlation coefficient of 0.90, indicating that the soluble component had more influence than the particulate in the CODLB. On the contrary, Karahan et al. (2008) found that the CODLB in the tannery effluent was strongly influenced by the particulate fraction, and by the trivalent chromium that inhibited the hydrolysis of its components. The difference in these results is attributable mainly to the pre-treatment to which the effluent was subjected, which removed chromium and the fraction of sedimentary solids (Kristensen et al., 1992; Orhon et al., 1999a).

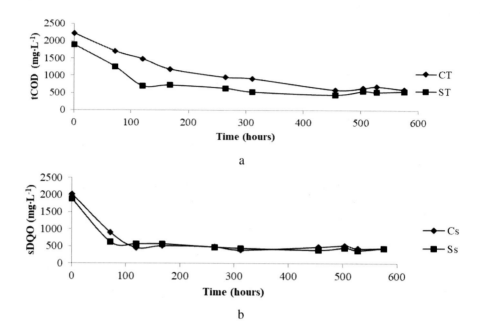

Figure 3. Determination of the chemical demand for particulate and soluble oxygen product of the microbial metabolism (Xp and Sp) in the effluent of the tannery. CT: Crude tCOD, Cs: Crude sCOD, ST: tCOD filtered, Ss: sCOD filtered.

Finally, in the study of the components of the COD, the soluble and particulate residual production of the metabolism of microorganisms was determined for the wastewater in the pond (Xp and Sp). It was found that 5.9 ± 1.4% of the tCOD would be generated as COD, remaining at the end of any biological process (Figure 3), being all of a particulate nature. While the soluble fraction was negligible, coinciding with that obtained by Germirli et al. (1991). This percentage is significant since it represents between 82.3 and 148.1 mg COD·L⁻¹ remaining at the end of the biological process, in addition to the inert soluble COD that remains, and which is not removable during the biological process. In this regard, researchers such as Kabdasli et al. (1993) and Orhon et al. (1999b) reported for tannery effluents, residual metabolic products between 3.0 and 4.5% (32 and 84 mg·L⁻¹) and of 5.1% (116.3 mg·L⁻¹), respectively.

Several researchers agree that Xp and Sp are important due to the nature of the industrial effluent (Germili et al., 1991; Orhon et al., 1994 and 1999b;

Orhon et al., 2009). Likewise, Bousier et al. (2005) pointed out that the inert soluble fraction and the inert particulate of the COD, pass without alteration through the treatment system, while a part of the biodegradable fraction is converted into a particulate microbial product (Xp), coming from the cellular lysis of the biomass during the experiments.

During the determination of the values of Xp and Sp, it was found that the reactors reached a stability of biological activities in an average time of 26.5 days, and at that moment, the concentrations of COD necessary to estimate Xp and Sp were determined graphically (Table 5). In this regard, Kabdasli et al. (1993) applying this procedure, achieved stabilization in 19 days for mixed tannery effluents. Additionally, Bousier et al. (2005) proposed that to verify that all biological activity had been completed, the stoichiometric coefficient Yxp (theoretical Yxp Experim<Yxp) should be calculated, which was found to be 0.07 ± 0.02 mg·mg^{-1} for this investigation, lower than the theoretical value of 0.2 reported by Orhon et al. (1999a), confirming that the biological process had been completed in its entirety.

Table 5. Initial and final COD[1] values for determination Xp and Sp

COD (mg·L^{-1})	Reactor 1 (unfiltered wastewater)	Reactor 2 (filtered wastewater)
tCOD initial	$2,225.6 \pm 410.7$	$1,898.5 \pm 464.9$
sCOD initial	$2,030.1 \pm 323.3$	$1,889.1 \pm 265.3$
tCOD final	600.0 ± 86.1	550.0 ± 53.1
sCODfinal	450.0 ± 71.4	450.0 ± 88.6

[1]COD concentrations taken from Figure 3. tCOD: total COD, sCOD: soluble COD.

NTK and N-NH$_4^+$ Behavior

It was observed that the concentrations of TKN and N-NH$_4^+$ of the wastewater in the pond presented a good association, with a R^2 of 0.8299 (Figure 4). Also, the low coefficients of variations shown in most cases in Table 3 indicate that only small changes occurred in the nitrogen concentrations during this research.

The concentration of chromium during the dry season was 6.3 mg·L^{-1} (Figure 5a). For the rainy period, the concentration of the metal was on average 3.8 mg·L^{-1}. Chromium concentration in the pond was closely related to pH, therefore large part of chromium was precipitated at the bottom of the pond. Loayza-Pérez (2006) found that the lower chromium solubility is reached between 7 and 8 units of pH, a range very close to the average pH of the wasteswater (Table 3).

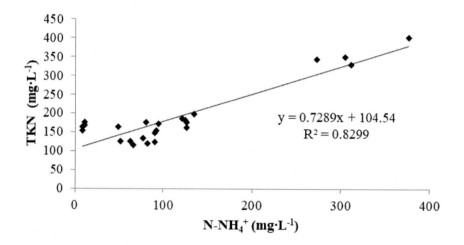

Figure 4. Regression analysis between ammoniacal nitrogen (N-NH$_4^+$) and total Kjeldahl nitrogen (TKN) of the wastewater contained in the pond.

The coefficients of variation (CV) ranged from 0 to 0.5 for variables such as pH, TSS, chromium, TKN, N-NH$_4^+$ and COD (Table 3), demonstrating the low variability of the concentration of contaminants in the effluent during the time and weather conditions. This provided an affirmation that the wastewater taken from the pond had reached an important stability during the time it was stored. This characteristic was important for selecting this effluent for study in this research.

It is important to highlight that the storage pond served as a homogenization and stabilization tank, in which physical, chemical and biological processes occurred naturally, and this contributed to the stabilization of the wastewater characteristics. Kristensen et al. (1992) noted

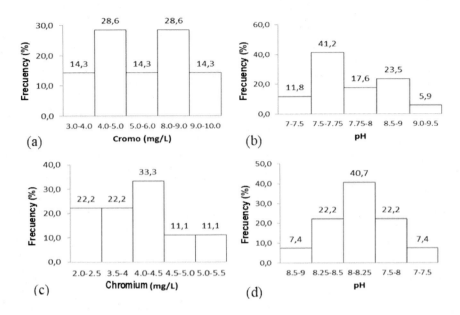

Figure 5. Frequency diagrams for the chromium concentrations and pH of the effluent storaged in the pond (a and c: drought period, b and d: rain period).

that the manipulation of wastewater by means of primary sedimentation or precipitation can dramatically change the organic profile of the effluents, and cause an important influence in the selection of the treatment process to apply to them. For this reason, an appropriate treatment process must be selected accordingly.

The characterization of tannery wastewater showed the great complexity of the matrix that forms it, due to the wide range of components that it has, such as the residues of raw materials (skins), proteins and fats, excess in the doses of the chemical reagents such as sulfides, sulfates and chlorides, as well as tanning agents such as chromium, dyes and surfactants (Munz et al., 2008; Durai and Rajasimman, 2011).

According to Munz et al. (2008) the selection of the treatment process for any effluent is strictly based on the distribution of the fractions of the COD, as well as the content of inhibitory compounds. For this reason, most of the characterization of the wastewater stored in the pond focused on the content of organic material; however, the matrix of the effluent also had an important content of inorganic material (Table 6).

Table 6. Inorganic elements present in the wastewater stored in the pond

Elements/Ions	Units	$\bar{X} \pm SD$	Discharge limits, Venezuelan law (MARNR,1995)
Metal			
Be	μg L^{-1}	2.8 ± 0.6	---
Na	mg L^{-1}	900.0 ± 180.0	---
Mg	mg L^{-1}	41.4 ± 5.10	---
Al	mg L^{-1}	1.8 ± 1.2	5.0
K	mg L^{-1}	66.2 ± 9.0	---
Ca	mg L^{-1}	21.0 ± 12.4	---
V	μg L^{-1}	94.7 ± 12.6	5,000[1]
Cr	mg L^{-1}	12.0 ± 6.0	2.0
Mn	μg L^{-1}	21.2 ± 20.4	2,000
Fe	mg L^{-1}	2.2 ± 1.3	10.0
Co	μg L^{-1}	0.0 ± 0.0	500.0
Ni	μg L^{-1}	102.9 ± 58.5	2.0[2]
Cu	μg L^{-1}	157.8 ± 63.4	1,000
Zn	μg L^{-1}	449.0 ± 102.4	5,000
As	μg L^{-1}	30.3 ± 6.8	500.0
Se	μg L^{-1}	86.1 ± 25.8	50.0
Mo	μg L^{-1}	18.4 ± 9.2	---
Ag	μg L^{-1}	107.6 ± 87.7	100.0
Cd	μg L^{-1}	24.3 ± 22.2	200.0
Sb	μg L^{-1}	0.0 ± 0.0	---
Ba	μg L^{-1}	286.8 ± 177.0	5,000
Hg	μg L^{-1}	12.0 ± 3.8	10.0
Tl	μg L^{-1}	0.0 ± 0.0	---
Pb	μg L^{-1}	57.8 ± 34.7	500.0
Th	μg L^{-1}	31.5 ± 0.4	---
U	μg L^{-1}	0.0 ± 0.0	---
No metal			
Cl$^-$	mg L^{-1}	8,549.5 ± 1,868.8	1,000
S^{2-}	mg L^{-1}	147.5 ± 6.5	0.5

[1]Regulated value for discharges to sewer networks, [2]Regulated value for discharges to marine-coastal environment and sewage networks.

The analysis of the effluent by means of ICP showed the presence of traces of various metals. Among them, however, only chromium, selenium, mercury and silver exceeded the limits established in the Venezuelan regulations for discharge to water bodies (MARNR, 1995). As previously mentioned, the presence of chromium is a common characteristic of wastewater from tanneries, and it was this heavy metal that exceeded the established limits by the greatest amount. The presence of the rest of the metals was only slightly higher than that allowed by environmental regulations, and was attributed to residues of chemical disinfectants added during the tanning process. On the other hand, the content of non-metallic ions, such as chlorides and sulfides (Table 6), that came from the reagents used in the leather production process, by far exceeded the limits established in Venezuelan legislation.

It should be noted that the results reported in this study correspond to the average concentrations of metals from 2009 to 2012 (Table 6); however, the chromium concentration was the most variable (3.8 and 24.5 mg·L^{-1}) was due to the characteristics of the production process. Therefore, in this investigation the concentration of this metal was reported at the beginning of each experiment.

These inorganic components are not removed by biological mechanisms; biotechnological processes such as biosorption, extracellular precipitation and uptake through biopolymers are required, which, when combined with physicochemical and biological processes, remove metals from the liquid phase (Gutiérrez et al., 2010). On the other hand, chlorides and sulfides can be removed by catalytic processes and ion exchange (Munz et al., 2008).

The combination of the conventional characterization, the fractionation of the COD and the determination of the inorganic content, allowed for the complete characterization of the effluent stored in the pond. It presented a pH of 8.1, suitable to apply a biological treatment, as well as the lowest concentration of chromium (6.3 mg·L^{-1}) and the lowest content of organic nitrogen. On the other hand, from the fractionation it was found that 57.4% of the tCOD, corresponding to the CODBT, could be removed by the action of microorganisms. Moreover, the fractionation gave indications of some of the conditions of operation of the biological system. CODNBp showed that sludge ages can vary from intermediate to high (10-20 days), which would favor the processes of nitrification and denitrification (Palma and Manga, 2005; González and Sadarriaga, 2008). Likewise, the fractionation of COD suggested that tannery wastewater treatment must combine biological and physicochemical processes.

Finally, the exhaustive and detailed characterization of the effluent stored in the pond showed that common parameters, such as TSS, COD, TKN, N-NH_4^+, chromium, sulfides and chlorides, do not comply with the discharge limits established in Venezuelan regulations (MARNR, 1995). For this reason, it was necessary to apply physical, chemical and/or biological treatment before discharging the wastewater into bodies of water.

DURATION OF THE CYCLE AND AERATION SEQUENCE FOR TANNERY EFFLUENTS USING A SBR

In this section, different oxic-anoxic aeration sequences have been incorporated to study the most suitable operational conditions for the biological treatment of the tannery effluent; these sequences have proven to be efficient in the biological removal of nutrients. A factorial analysis was carried out where two factors were tested: duration of the cycle and the aeration sequence. The cycle duration had 4 levels (8, 10, 12 and 24 h), while the aeration sequence had 3 levels (intermittent aeration, conventional nitrification-denitrification and oxic reaction). For this stage of the investigation, three SBRs were used that worked in parallel, all fed with the same tannery wastewater.

The characteristics of the raw wastewater from the storage pond are shown in Table 7. The average volumetric organic load (VOC) for each of the cycle durations tested was 1.9, 1.5, 1.3 and 0.6 $kg \cdot m^{-3} \cdot d^{-1}$ for 8, 10, 12 and 24 h, respectively. While the average loads of ammoniacal nitrogen were 0.08, 0.06, 0.05 and 0.03 $kg\ N\text{-}NH_4^+ \cdot m^{-3} \cdot d^{-1}$ for the same cycle periods. The hydraulic residence time (HRT) was 30 h with a volume exchange ratio (VER) of 40%, while the cellular retention time was maintained at approximately 15 d, to permit the establishment of nitrifying bacteria in the biomass (Ekama and Wentzel, 2008a; Carrasquero, 2011).

Table 7. Characterization of wastewater from the tannery

		$\overline{X} \pm SD$		
		SBR 1	**SBR 2**	**SBR 3**
Variable /SBR operation condition		**Intermittent aeration (Anox-Ox-Anox-Ox-Anox-Ox)**	**Conventional nitrification-desnitrification (Ox-Anox)**	**Only oxic phase (Ox)**
[1]tCOD(mg·L^{-1})		1,602.3 ± 7.3	1,557.3 ± 59.3	1,523.6 ± 33.9
[2]TKN(mg·L^{-1})		122.5 ± 5.3	115.5 ± 7.9	108,5 ± 12.0
[3]N-NH$_4^+$ (mg·L^{-1})		60.4 ± 4.4	66.5 ± 3.1	62.1 ± 6.6
pH		8.2 ± 0.3	8.2 ± 0.2	8.2 ± 0.3
[4]VOC (kg·m^{-3}·d^{-1})	8 h	1.9	1.8	1.8
	10 h	1.5	1.5	1.5
	12 h	1.3	1.2	1.2
	24 h	0.6	0.6	0.6
N-NH$_4^+$ load (kg·m^{-3}·d^{-1})	8 h	0.07	0.08	0.07
	10 h	0.06	0.06	0.06
	12 h	0.05	0.05	0.05
	24 h	0.02	0.03	0.02

[1]DQO: Total oxygen chemical demand, [2]TKN: Kjeldahl total nitrogen, [3]N-NH$_4$ +: Ammoniacal nitrogen, [4]VOC: Volumetric organic load. Ox: oxic, Anox: anoxic.

COD Behavior

The concentration of COD after treatment in the SBRs decreased consistently as the cycle duration increased from 8 to 12 h, decreasing from 951.4 to 831.9 mg COD·L^{-1} (Figure 6); however, when the 24 h cycle was tested, an increase in COD concentration at the reactor outlet was observed (937.1 mg COD·L^{-1}), This was attributable to the decay of the biomass during the endogenous metabolism phase, as a product of cell lysis, in which the cytoplasmic content of dead cells is released, represent an increase in the COD of the effluent (Orhon et al., 1989; Kabdasli et al., 1993; Garzón-Zúñiga, 2005).

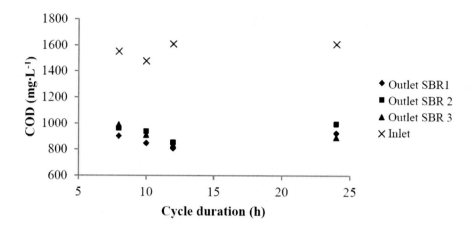

Figure 6. Behavior of the chemical oxygen demand (COD) in the SBRs for the four cycle durations studied.

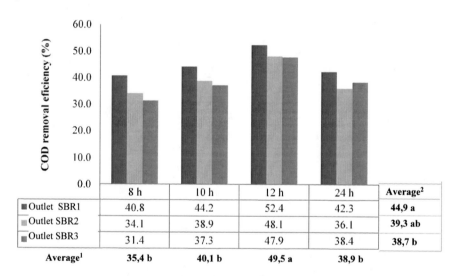

	8 h	10 h	12 h	24 h	Average[2]
Outlet SBR1	40.8	44.2	52.4	42.3	**44,9 a**
Outlet SBR2	34.1	38.9	48.1	36.1	**39,3 ab**
Outlet SBR3	31.4	37.3	47.9	38.4	**38,7 b**
Average[1]	**35,4 b**	**40,1 b**	**49,5 a**	**38,9 b**	

Figure 7. Efficiency of chemical oxygen demand (COD) removal for four cycle times and three aeration sequences. Removal = $(COD_{inlet}-COD_{outlet})/COD_{inlet} * 100$ [1]COD removal average for each cycle duration. [2]COD removal average for each aeration sequence. X axis: Cycle duration (h).

The highest COD removal was obtained for the 12 h cycle, reaching average removals of 49.5% (Figure 7), this being significantly higher than in the other cycle durations (P≤0.05). No statistical differences were observed between them when cycles of 8, 10 and 24 h were tested (P>0.05). With regard to the sequences of aeration, it was found that the sequence Anox/Ox/Anox/Ox/Anox/Ox, used in SBR 1 (intermittent aeration), was the best to remove COD, with 44.9% efficiency. On the other hand, the reactor that worked only with the oxic phase (SBR 3) was the one that statistically removed the lowest amount of COD (38.7%). The Anox/Ox sequence used in SBR 2 (conventional nitrification-denitrification) achieved 39.3% removals, with no significant difference between the other two sequences (P>0.05).

The minimum COD concentration reached in the SBR 1 after the treatment was approximately 800 mg·L^{-1}, being a high concentration value, indicative of the presence of non-biodegradable material in the tannery's effluent (Kabdasli et al., 1993; Munz et al., 2008; Mekonnen and Leta, 2011).

From the fractionation of the COD it was found that 42.6% of the total COD of the tannery wastewater is inert, equivalent to 682.5 mg COD·L^{-1}. Most of this remaining COD is an inert soluble nature, in addition to that generated by microbial metabolism, representing at least 465 mg·L^{-1} of residual COD, which remains unchanged during biological treatment and it is measured as COD in the discharge of the reactors. Ganesh et al. (2006) concluded that the problem associated with the high concentrations of COD found at the end of a biological treatment in effluents is an intrinsinc characteristic of them. They suggest as a solution, the substitution of the chemical reagents used in the productive process for those which cause less environmental impact.

The analysis of the factorial arrangement for COD removal showed that there was no interaction between the two factors studied (P>0.05), for all the aeration sequences tested, the best cycle time was 12 h. On the other hand, the best aeration sequence was the one that intercalated short anoxic and oxic periods used in the SBR 1 (intermittent aeration), being independent of the duration of the cycle.

Nitrogen Behavior

Concentration of ammoniacal nitrogen at the outlet of the SBRs decreased continuously as the cycles became longer (Figure 8). In particular, the SBR 3 showed the greatest removal of $N-NH_4^+$, because the duration of the oxic phase was the longest, favoring the nitrification process, which is a process that takes place under aerobic conditions (Metcalf & Eddy, 1995; Garzón-Zúñiga, 2005).

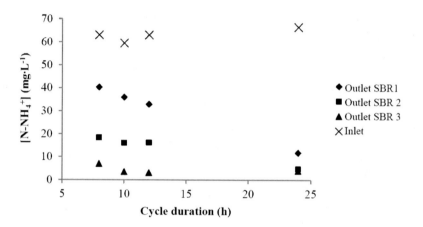

Figure 8. Behavior of ammonia nitrogen ($N-NH_4^+$) in the SBRs for the four cycle durations studied.

In Figure 9 it can be seen that the highest efficiency of nitrification was obtained for the SBR 3 (92.9%). On the other hand, the anoxic-oxic sequence used in the SBR 2 showed an intermediate efficiency for the removal of $N-NH_4^+$ (72.1%), while the sequence Anox/Ox/Anox/Ox/Anox/Ox was statistically the lowest of all, achieving only 49.6% transformation of ammoniacal nitrogen to nitrites and nitrates ($P \leq 0.05$). Regarding the duration of the cycle, it was found that the longest cycle (24 h) was the one that produced the highest nitrification efficiency, achieving 88.8%. For the rest of the tested times there was no significant difference between them ($P > 0.05$), but they were statistically lower than the efficiency of the 24 h cycles ($P \leq 0.05$).

The statistical analysis showed there was interaction between the two studied factors (cycle duration and aeration sequences), negatively affecting the removal of ammoniacal nitrogen in the SBR 1 for all cycle times tested (Figure 9), while for the other sequences and all cycle times there was no significant difference in nitrification efficiency ($P>0.05$).

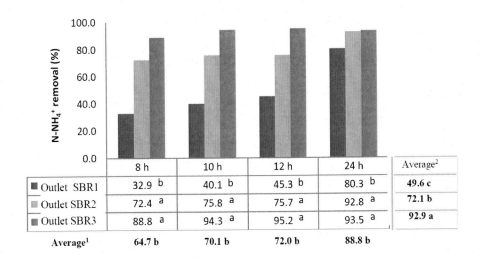

	8 h	10 h	12 h	24 h	Average[2]
■ Outlet SBR1	32.9 b	40.1 b	45.3 b	80.3 b	49.6 c
▨ Outlet SBR2	72.4 a	75.8 a	75.7 a	92.8 a	72.1 b
▨ Outlet SBR3	88.8 a	94.3 a	95.2 a	93.5 a	92.9 a
Average[1]	64.7 b	70.1 b	72.0 b	88.8 b	

Figure 9. Removal efficiency of ammoniacal nitrogen ($N-NH_4^+$) for four cycle times and three aeration sequences. [1]$N-NH_4^+$ removal average for each cycle duration, [2]$N-NH_4^+$ removal average for each aeration sequence. X axis: Cycle duration (h).

It is important to note that for the SBR 1 and SBR 2 the duration of the oxic phase was the same, the only difference was it was divided into three parts for SBR 1. This result indicated that nitrifying microorganisms could not be fully established in SBR 1, and it was the first to be affected when the anoxic phases were intercalated. Similar observations were reported by Farabegoli et al. (2004) and by Lefevbre et al. (2005) when they studied tannery effluents, and they observed that nitrifying microorganisms are the first to be affected, decreasing their activity, when they are exposed to certain concentrations of inhibiting substances, such as chromium and sodium chloride, respectively; this effect could, in certain cases, become irreversible.

It should be noted that Figure 9 shows that an upward trend was maintained in the removal of ammoniacal nitrogen for the SBR 1 as the duration of the cycle increased, which indicated that the inhibitory process was not irreversible, on the contrary, the group of nitrifiers apparently required more cycle time (longer oxic phases) to adapt to intermittent aeration conditions.

For the selection of the duration of the cycle, the behavior of the reactors was evaluated and it was found that the most suitable time was 12 h, during which time the highest COD removal efficiencies ($\approx$50%) and intermediate removals of ammonia nitrogen (72.0%) were achieved. It was observed that the nitrification efficiency can be increased by allowing the nitrifying microorganisms to be properly established in the biological system. In addition, using 12 h cycles, twice the volume of residual water in the SBR would be processed per day compared, to the 24 h cycles. The cycles of 8 h and 10 h reached lower COD removals, and showed no improvement in nitrification in relation to the 12 h cycle. Therefore, they did not represent adequate treatments for the tannery effluent.

Ganesh et al. (2006) treated effluents of tanneries in an aerobic SBR and determined that the optimum cycle time was 12 h, achieving removals of 81% of COD and 85% of $N-NH_4^+$, when the volumetric organic load varied between 1.9 and 2.1 $kg \cdot m^{-3} \cdot d^{-1}$ and the hydraulic residence time (HRT) was 1 d. Despite having agreed that the optimal cycle time is 12 h, and using a closer HRT (1.25 d), it is important to highlight that the volumetric organic load used was slightly lower (1.3 $kg \cdot m^{-3} \cdot d^{-1}$) that achieved by these researchers, and also the efficiency of the system was lower (49.5%). These results can be attributed to the effluent used by Ganesh et al. (2006) which presented better biodegradability characteristics, containing on average 19% non-biodegradable material versus 42.6% that of the effluent stored in the pond.

Mekonnen and Leta (2011), using wastewater from tanneries in an SBR, determined that for the longer cycle (24 h), the removal of COD and total nitrogen (TN) was higher. However, making a cost-time-benefit analysis, they determined that a duration of 8 h, with 32 h of hydraulic residence time (TRH) and 2.3 $kg \cdot m^{-3} \cdot d^{-1}$ of VOC, corresponded to the optimal conditions of operation of the SBR. In these cases there were intermediate removals of

COD (85%) and TN (39%), and a greater efficiency of removal of mass of pollutants per unit of time.

Lefebvre et al. (2005) achieved the optimal conditions to treat soak liquor effluent, 0.6 kg·m^{-3}·d^{-1}, and 5 days of HRT, achieving removals of 95 and 96% of COD and TKN, respectively. However, such extensive HRT and low VOCs counteract the benefits of using SBR as a treatment technology (Ganesh et al., 2006). This aspect is of importance in this investigation, because the optimal VOC and HRT results were found among those reported by Lefebvre et al. (2005) and Ganesh et al. (2006).

The aeration sequence showed opposite behaviors, according to the purpose of the biological treatment. The best sequence for the removal of COD was intermittent aeration (SBR 1), while greater removal of N-NH$_4^+$ was achieved in the SBR 3 working under aerobic conditions. The sequence of SBR 2 had behavior intermediate between the other two treatments, achieving similar removals of COD as SBR 1, and removal of N-NH$_4^+$ less than SBR 3. Therefore, the selection criteria of the aeration sequence will depend on the objective of the wastewater treatment.

Finally, the most appropriate cycle time for the removal of COD and N-NH$_4^+$ was 12 h for the effluent stored in the pond, regardless of the aeration sequence used. The operation conditions of the SBR were 30 h of TRH, 1.3 kg·m^{-3}·d^{-1} of VOC and Θ_C of 15 days, under these conditions removals of 49.5% of COD and 72% of N-NH$_4^+$ were achieved.

REMOVAL OF NUTRIENTS FROM TANNERY WASTEWATER USING A SBR

The removal efficiency of organic matter and nutrients was evaluated in sequential batch reactors (SBR), operated under different aeration sequences and filling times, in order to determine the best conditions for the operation of biological treatment for removing COD and nitrogen.

Operational Conditions of the SBRs for Each Treatment Studied

The working conditions for the SBRs were similar for the twelve treatments tested, although the experiments were carried out at different times of the year (Table 8). The slight variations between the treatments are attributed to the inherent variability of the wastewater, as mentioned on Tables 1 and 3. The operational conditions in the SBRs were cycle of 12 h, HRT of 30 h and sludge age of 15 d.

The pH was within the recommended range for the growth of microorganisms in a biological system (Grunditz and Dalhammar, 2001), with values between 7.9 and 8.3 units for raw wastewater; after biological treatment in the SBR, an increase was observed, with pH values between 8.1 and 8.9 units (Table 8). It is important to note that the wastewater forms a buffer that maintains the pH with little variation. This behavior coincides with observation made by Garbagnati et al. (2005), who reported that tannery wastewaters have a buffer capacity which prevents the determination of the typical pH and alkalinity variations during biological treatments in an SBR. Similarly, Li and Irvin (2007) noted that the range of pH variation is a direct function of the buffer capacity of the liquid.

On the other hand, the pH of the raw wastewater contributed to the low concentration of total chromium in the liquid phase (13.4 mg $Cr \cdot L^{-1}$), because the solubility of chromium in solutions with neutral pH is only 0.08 $mg \cdot L^{-1}$. As the residual water becomes slightly more alkaline (pH of 8.5) the solubility reaches its maximum, which is only 0.5 $mg \cdot L^{-1}$ (Cado, 1996), so that the highest concentration of total chromium precipitated, accumulating in the sediment at the bottom of the storage pond. Farabegoli et al. (2004) determined that the inhibition of biological processes was greater for nitrification than for denitrification, because the toxicity of the metal has a greater effect on the nitrifying microorganisms. However, the inhibition was observed at concentrations much higher than it reported in the present research (120 mg $Cr \cdot L^{-1}$).

Table 8. Operational parameters in the SBRs for the treatments tested ($\overline{X} \pm SD$)

Treatments	pH		T^1	VOC^2	CNT^3	$TSSML^4$	$VSSML^5$
	Inlet	Outlet	(°C)	(kg·m⁻³·d⁻¹)	(kg·m⁻³·d⁻¹)	(mg·L⁻¹)	(mg·L⁻¹)
AI_R[6]	7.9 ± 0.55	8.7 ± 0.16	27.5 ± 0.3	1.08 ± 0.02	0.14 ± 0.03	7,980.0 ± 1,448.5	4,982.9 ± 1,050.5
AI_L[7]	8.1 ± 0.56	8.5 ± 0.15	26.3 ± 0.9	1.26 ± 0.20	0.13 ± 0.01	7,270.0 ± 730.8	4,353.3 ± 449.7
AI_E[8]	8.3 ± 0.29	8.4 ± 0.09	26.4 ± 0.8	1.26 ± 0.14	0.11 ± 0.01	6,778.0 ± 637.5	3,623.0 ± 170.4
CND_R[9]	7.9 ± 0.55	8.7 ± 0.10	27.3 ± 0.4	1.09 ± 0.02	0.14 ± 0.03	9,720.0 ± 1,581.3	6,014.3 ± 1,474.8
CND_L[10]	8.1 ± 0.56	8.5 ± 0.26	26.5 ± 0.5	1.17 ± 0.02	0.13 ± 0.01	6,923.3 ± 687.0	4,230.0 ± 356.8
CND_E[11]	8.2 ± 0.4	8.3 ± 0.15	26.5 ± 0.8	1.24 ± 0.14	0.11 ± 0.01	7,080.0 ± 374.8	3,573.0 ± 369.7
NDS_R[12]	8.3 ± 0.13	8.3 ± 0.07	26.4 ± 0.7	1.13 ± 0.13	0.13 ± 0.01	8,167.0 ± 807.1	4,043.0 ± 295.7
NDS_L[13]	7.9 ± 0.43	8.1 ± 0.19	25.7 ± 0.4	0.94 ± 0.08	0.11 ± 0.01	6,065.0 ± 233.3	3,145.0 ± 148.5
NDS_E[14]	7.6 ± 0.39	8.1 ± 0.06	24.6 ± 0.4	1.10 ± 0.12	0.10 ± 0.00	8,658.0 ± 1,047.0	4,158.0 ± 530.3
PD_R[15]	8.2 ± 0.03	8.9 ± 0.26	26.4 ± 0.6	1.19 ± 0.17	0.13 ± 0.01	7,446.7 ± 639.6	3,993.0 ± 153.1
PD_L[16]	7.9 ± 0.42	8.7 ± 0.28	25.6 ± 0.5	0.94 ± 0.08	0.12 ± 0.01	7,825.0 ± 745.9	4,140.0 ± 113.1
PD_E[17]	7.5 ± 0.38	8.4 ± 0.08	25.2 ± 0.6	1.08 ± 0.02	0.10 ± 0.00	9,420.0 ± 1,100.0	4,582.5 ± 618.5

[1] T: Temperature, [2] VOC: Volumetric organic load, [3] CNT: Total nitrogen load, [4] TSSML: Total suspended solids of the mixed liquor, [5] VSSML: Volatile suspended solids of the mixed liquor, [6] AI_R: Intermittent aeration, fast filling, [7] AI_L: Intermittent aeration, slow filling, [8] AI_E: Intermittent aeration, stepwise filling, [9] NDC_R: Nitrification, conventional denitrification, fast filling, [10] CND_L: Conventional nitrification, conventional denitrification, slow filling, [11] CND_E: Conventional nitrification, conventional denitrification, stepwise filling, [12] NDS_R: simultaneous nitrification denitrification, fast filling, [13] NDS_L: simultaneous nitrification denitrification, slow filling, [14] NDS_E: simultaneous nitrification denitrification, stepwise filling, [15] PD_R: Predesnitrification, fast filling, [16] PD_L: Predesnitrification, slow filling, [17] PD_E: Predesnitrification, filling by stages.

**Table 9. Inorganic elements in the tannery wastewater
during biological treatment**

Elements	Units	SBR Inlet	SBR Outlet
Metals		$(\bar{X} \pm SD)$	
Be	$\mu g \cdot L^{-1}$	10.8 ± 0.03	10.9 ± 0.0
Na	$mg \cdot L^{-1}$	3,600.3 ± 117.7	3,826.0 ± 254.5
Mg	$mg \cdot L^{-1}$	43.4 ± 1.5	13.3 ± 9.0
Al	$\mu g \cdot L^{-1}$	3.7 ± 0.09	4.7 ± 0.9
K	$mg \cdot L^{-1}$	64.9 ± 0.9	60.1 ± 4.4
Ca	$mg \cdot L^{-1}$	41.8 ± 5.2	18.7 ± 2.8
V	$\mu g \cdot L^{-1}$	138.6 ± 6.0	116.5 ± 3.7
Cr	$mg \cdot L^{-1}$	13.4 ± 1.71	8.9 ± 5.8
Mn	$\mu g \cdot L^{-1}$	53.6 ± 9.3	34.9 ± 7.5
Fe	$mg \cdot L^{-1}$	6.3 ± 0.8	7.0 ± 0.6
Co	$\mu g \cdot L^{-1}$	0.0 ± 0.0	0.0 ± 0.0
Ni	$\mu g \cdot L^{-1}$	129.3 ± 39.3	83.5 ± 27.4
Cu	$\mu g \cdot L^{-1}$	216.2 ± 19.4	210.2 ± 35.3
Zn	$\mu g \cdot L^{-1}$	505.6 ± 125.4	551.0 ± 192.4
As	$\mu g \cdot L^{-1}$	40.0 ± 1.2	24.5 ± 1.6
Se	$\mu g \cdot L^{-1}$	0.1 ± 0.3	0.0 ± 0.0
Mo	$\mu g \cdot L^{-1}$	13.9 ± 1.9	14.7 ± 0.7
Ag	$\mu g \cdot L^{-1}$	0.0 ± 0.0	74.0 ± 148.0
Cd	$\mu g \cdot L^{-1}$	0.0 ± 0.0	0.0 ± 0.0
Sb	$\mu g \cdot L^{-1}$	0.0 ± 0.0	0.0 ± 0.0
Ba	$\mu g \cdot L^{-1}$	871.4 ± 98.8	755.1 ± 42.3
Hg	$\mu g \cdot L^{-1}$	10.5 ± 0.2	4.8 ± 0.03
Tl	$\mu g \cdot L^{-1}$	0.0 ± 0.0	0.0 ± 0.0
Pb	$\mu g \cdot L^{-1}$	26.2 ± 20.5	23.3 ± 7.5
Th	$\mu g \cdot L^{-1}$	31.2 ± 0.2	31.3 ± 0.1
U	$\mu g \cdot L^{-1}$	0.0 ± 0.0	0.0 ± 0.0

The volumetric organic loads (VOC) among the 12 treatments studied, ranged between 0.94 - 1.26 $kg \cdot m^{-3} \cdot d^{-1}$, being higher than 0.6 $kg \cdot m^{-3} \cdot d^{-1}$ reported by Lefebvre et al. (2005) and inferior to those used by Ganesh et al. (2006) and Mekonnen and Leta (2011), who treated effluents from tanneries in SBR with a VOC of 2.0 and 2.3 $kg \cdot m^{-3} \cdot d^{-1}$, respectively and hydraulic retention times of 5, 1 and 1.3 d. The effluents used by the researchers mentioned above had higher COD and total nitrogen concentrations than those obtained in the present investigation.

During the evaluation of the removal of nutrients, the determination of the content of inorganic material (metals) was made at the beginning and at the end of the biological treatment (Table 9). A decrease in the concentration of chromium, mercury, barium, arsenic, nickel, magnesium, calcium and vanadium was observed, presumably caused by the chemical precipitation due to the slightly basic pH of the mixed liquor (Loayza, 2006). This could also happen by the process of biosorption, which consists of the extraction of the metal by a physicochemical interaction in the cellular surface of the microorganisms, attracting them from the solution and linking them to the biomass (Cañizares - Villanueva, 2000).

Efficiency of the SBR in the Removal of COD

Table 10 shows that the efficiency of the SBR for removal of COD was better than the treatments using intermittent aeration (AI), pre-denitrification (PD) and conventional sequence nitrification-denitrification (CND), achieving removals between 53, 2 and 55.1% of COD, without a statistical difference between them (P>0.05). However, each strategy achieved that removal efficiency when different filling times were applied, that is, there was interaction between the two factors studied, being the factor "filling time," statistically responsible for the removal efficiency.

The lowest COD removal occurred in the reactor that used CND and fast filling, achieving efficiencies of only 38.2%, being statistically different to AI_L, PD_R and CND_E. For the rest of the treatments studied there were no significant differences in the efficiency of the biological system, and the range of removal was between 40.4 and 52.2% (P>0.05).

The inlet tCOD to the SBR ranged between 1,174.8 and 1,576.8 mg·L^{-1}. The difference between these concentrations is due to the fact that the tannery effluents are variable and complex, which increases the difficulty in applying an adequate wastewater treatment. The variability is the result of the difference in the volumes generated from any of the fractions (soak liquor, tanning and/or dyeing), as well as the different chemical reagents used (Kabdasli et al., 1993; Galisteo et al., 2002).

Table 10. Efficiency of COD removal for the treatments studied

Treatment	Total chemical oxygen demand (DQO$_t$)			
	SBR inlet (mg·L^{-1})	SBR outlet (mg·L^{-1})	Total remotion* (%)	SBR efficienciy[13] (%)
AI$_L$[1]	1,576.8 ± 247.9	702.2 ± 191.3	55.1 a	98.5 a
PD$_R$[2]	1,485.8 ± 213.2	694.3 ± 85.2	53.4 a	96.3 a
CND$_E$[3]	1,549.0 ± 179.9	724.9 ± 151.6	53.2 a	93.3 ab
CND$_L$[4]	1,457.4 ± 20.6	696.7 ± 105.5	52.2 ab	92.3 ab
NDS$_R$[5]	1,412.3 ± 168.7	725.8 ± 387.5	48.6 abc	91.6 abc
NDS$_E$[6]	1,377.2 ± 143.9	712.0 ± 35.9	48.0 abcd	91.3 abc
AI$_E$[7]	1,570.4 ± 179.9	815.4 ± 162.8	48.1 abcd	88.3 abc
NDS$_L$[8]	1,174.8 ± 105.7	618.0 ± 122.1	47.4 abcd	90.4 abc
PD$_E$[9]	1,353.9 ± 136.4	714.6 ± 71.2	47.1 abcd	90.2 abc
AI$_R$[10]	1,354.8 ± 27.5	781.9 ± 73.3	42.3 bcd	85.3 abc
CND$_R$[11]	1,363.7 ± 27.5	811.7 ± 94.9	40.4 cd	81.2 c
PD$_L$[12]	1,174.8 ± 105.7	722.0 ± 116.0	38.5 d	83.4 bc

[1]AI$_L$: Intermittent aeration, slow filling, [2]PD$_r$: Predesnitrification, fast filling, [3]CND$_E$: Conventional nitrification denitrification, stepwise filling, [4]NDC$_L$: Conventional nitrification denitrification, slow filling, [5]NDS$_R$: Simultaneous nitrification denitrification, fast filling, [6]NDS$_E$: Simultaneous nitrification denitrification, stepwise filling, [7]AI$_E$: Intermittent aeration, stepwise filling, [8]NDS$_L$: Simultaneous nitrification denitrification slow filling, [9]PD$_E$: Predesnitrification, stepwise filling, [10]AI$_R$: Intermittent aeration, fast filling, [11]CND$_R$: Conventional nitrification denitrification, slow filling, [12]PD$_L$: Predesnitrification, slow filling, [13]Efficiency of the SBR: Considering the removal only of the biodegradable material.
*Mean followed by different letters in each column indicates significant differences according to the Tukey test (P≤0.05).

Considering that the effluent stored in the pond had a non-biodegradable content that ranged between 36 and 42%, it was decided to calculate the removal efficiency of the SBR in terms of only the biodegradable COD (Table 10). It was found that the best treatments, mentioned above, AI$_L$, PD$_R$ and CND$_E$, removed 98.5, 96.3 and 93.3% of the biodegradable COD, respectively, without a statistical difference between them (P>0.05), and they were higher than CND$_R$ that removed 81.2% (P≤0.05). These results showed that the SBR was efficient in the removal of COD, coinciding with the results obtained by other researchers who also used an SBR to treat effluents, and obtained COD removals between 75 and 95% (Vidal et al., 2004; Lefebvre et al., 2005; Ganesh et al., 2006; Ryu et al., 2007; Mekonnen and Leta, 2011).

The efficiency of the treatments that achieved the greatest COD removal had a good correspondence with the content of biodegradable material, obtained according to the fractionation of the COD (57%). The remaining COD (702.2, 694.3 and 724.9 mg·L^{-1}) corresponds to non-biodegradable material (inert soluble COD and COD product of cellular metabolism). This remaining portion was higher by approximately 50% of the discharge limit, than allowed by Venezuelan environmental regulations (MARNR, 1985), and therefore, must be removed by means other than biological (Orhon et al., 1999a; Dosta et al., 2008).

Efficiency of SBR in Nitrification and Denitrification

Nitrification efficiency in the SBR was above 90% for most of the treatments tested (AI$_L$, CND$_L$, AI$_R$, PD$_L$, PD$_R$, CND$_R$, NDS$_R$) with removals between 91.5 and 97.0% (P>0.05). The smallest removals occurred with fillings in stages, and the NDS$_L$ treatment, in which the removal was slightly higher than 70%. For the rest of the treatments, there was no significant difference between them (Table 11). The statistical analysis showed that the two factors (filling time and aeration sequence) had an effect on the nitrification efficiency.

It is important to highlight that the concentration of ammoniacal nitrogen at the outlet of the SBR in most of the treatments was less than 10 mg·L^{-1}, it shows an adequate nitrification capacity in the SBR. The largest ammoniacal nitrogen removals were achieved with the AI and PD sequences, being statistically superior to that achieved with the NDS strategy (P≤0.05). This behavior confirms that the presence of trivalent chromium in a concentration of 13.4 mg·L^{-1} did not affect the biological process of nitrification in a significant way; these being attributable to the lower registered nitrification efficiencies in operational parameters such as duration, sequence of phases and/or cycles, and forms and times of filling the SBR, among others.

Table 11. Nitrification efficiency in SBRs for the treatments studied

Treatment	N-NH$_4^+$			Nitrification rates[*]
	SBR inlet (mg·L^{-1})	SBR outlet (mg·L^{-1})	Total removal* (%)	(mg N-NO$_3^-$· mg SSV^{-1}·d^{-1})
AI$_L$[1]	91.9 ± 1.8	2.8 ± 1.0	97.0 a	0.0794 a
CND$_L$[2]	91.9 ± 1.8	4.5 ± 3.8	95.1 a	0.0410 c
AI$_R$[3]	68.7 ± 10.8	4.4 ± 1.0	93.7 a	0.0160 gh
PD$_L$[4]	70.0 ± 4.0	4.6 ± 1.8	93.4 a	0.0320 d
PD$_R$[5]	84.0 ± 7.0	5.7 ± 1.5	93.2 a	0.0170 gh
CND$_R$[6]	66.8 ± 7.2	5.7 ± 2.1	91.5 a	0.0080 i
NDS$_R$[7]	84.0 ± 7.0	7.5 ± 3.6	91.5 a	0.0150 h
AI$_E$[8]	87.5 ± 7.6	8.4 ± 1.1	86.4 ab	0.0457 b
CND$_E$[9]	86.6 ± 9.2	15.6 ± 2.7	79.4 bc	0.0309 d
NDS$_E$[10]	63.9 ± 1.8	16.2 ± 1.5	74.6 c	0.0240 ef
PD$_E$[11]	63.9 ± 1.8	17.6 ± 4.7	72.3 c	0.0260 e
NDS$_L$[12]	70.0 ± 4.0	20.3 ± 4.8	71.1 c	0.0202 fg

[1]AI$_L$: Intermittent aeration, slow filling, [2]CND$_L$: Conventional nitrification denitrification, slow filling, [3]AI$_R$: Intermittent aeration, fast filling, [4]PD$_L$: Predesnitrification, slow filling, [5]PD$_R$: Predesnitrification, fast filling, [6]CND$_R$: Conventional nitrification denitrification, fast filling, [7]NDS$_R$: Simultaneous nitrification denitrification, fast filling, [8]AI$_E$: Intermittent aeration, stepwise filling, [9]CND$_E$: Conventional nitrification denitrification, stepwise filling, [10]NDS$_E$: Simultaneous nitrification denitrification, stepwise filling, [11]PD$_E$: Predesnitrification, stepwise filling, [12]NDS$_L$: Simultaneous nitrification denitrification, filling in stages.
*Mean followed by different letters in each column indicates significant differences according to the Tukey test (P≤0.05).

It was observed the less efficient treatments corresponded to the strategy of filling in stages (CND$_E$, NDS$_E$, PD$_E$). This behavior was attributed to the fact that the intermittent filling added fresh residual water in different periods of the cycle in the SBR, and the lapses that were given for nitrification in the last subcycle were not enough, leaving ammoniacal nitrogen without nitrification at the end of the complete cycle. Therefore, the strategy of filling by stages was the one that removed the least amount of N-NH$_4^+$ (P≤0.05).

On the other hand, the behavior of the specific nitrification rates showed that the intermittent aeration sequence reached higher values than the CND and PD sequences, which were superior to the NDS (P≤0.05). Regarding the filling time, the specific nitrification rate during the slow filling was significantly higher than that of the filling by stages, and these in turn they were higher than the nitrification rate of the fast filling (Table 11).

For all treatments, the specific nitrification rates, shown in Table 10, were higher than that obtained by Ganesh et al. (2006) who reported 0.0069 mg·mg^{-1}·d^{-1} for effluents treated in an SBR, with NDS filling and the rest of the cycle in oxic conditions (8 h). While they were inferior to that reported by Farabegoli et al. (2004) who determined it at 0.078 mg·mg^{-1}·d^{-1} in effluents without the presence of trivalent chromium and sulfides, with the exception of the AI$_L$ treatment for which the rate was similar (0.0794).

The differences between the specific velocities, apart from the characteristics of the wastewater, are due to the fact that the rate is inversely proportional to the concentration of volatile suspended solids in the liquor mixture of each SBR, and therefore, the content and activity. The fraction of nitrifying microorganisms was specific to each biomass used.

The specific nitrification rate had a good correspondence with the removal of ammoniacal nitrogen for the AI$_L$ treatment, both values being significantly higher than the rest of the treatments tested (P≤0.05), which implies that the activity of the nitrifying biomass was high, and therefore duration of oxic phase can be reduced to optimize the system. However, this behavior was not maintained for all the treatments applied, because the efficiency of nitrification depends directly on the duration of the oxic phase and not on the nitrification rate (speed with which the nitrifying microorganisms manage to oxidize the ammonium).

Because of this, the variability obtained in the nitrification rates for each treatment had little impact on the nitrification efficiency of the system, indicating that the duration of the oxic phase was sufficiently long to guarantee the greater oxidation of the ammonia nitrogen in the majority of the treatments studied. The exception to this behavior were those treatments in which stage filling was used, because the ammonia removal was significantly lower than the rest of the treatments, implying that rates did not represent sufficient speeds for oxidizing the N-NH$_4^+$ added in the last stage of the filling step, which produced concentrations of $N-NH_4^+$ higher than 10 mg·L^{-1} in the SBR effluent (Table 11).

The independence of the nitrification rate with respect to the efficiency of this process was evidenced in the PD_R and PD_L treatments, in which, although the nitrification speed was low (Table 11), the removal eficiency of ammoniacal nitrogen occurred with a 93.2- 93.4%, demonstrating that the aerated fraction, which relates the duration of the oxic phase and the reaction time in the SBR, was sufficient to overcome the problem that could have caused the low nitrification rate. Insel et al. (2006) noted that the aerated fraction is an important parameter in the design of treatment systems based on strategies of intermittent aeration and pre-denitrification. They also indicated that empirically, in order to guarantee complete nitrification in domestic wastewaters, this ratio should be 0.5. However, to obtain adequate removals of ammoniacal nitrogen for the effluent stored in the pond, a greater aerated fraction was required during the cycles in the SBR ($\approx$0.64).

Regarding TKN removal, which includes the sum of the content of ammonia nitrogen and organic nitrogen, it was found that the best treatment was represented by the AI sequence with slow filling (73.7%), being significantly superior to PD_E and NDS_E (P≤0.05) in which 52.0 and 50.1% were removed, respectively (Table 12). The slow filling time was statistically superior to fast filling and this was greater than filling by stages (P≤0.05). On the other hand, the best sequence found was intermittent aeration, being superior to simultaneous nitrification-denitrification (P≤0.05). The other two aeration sequences tested showed no statistical difference, being similar to those mentioned above (P>0.05).

It can be seen that both the efficiency of nitrification and the removal of TKN were favored with slow fillings. Sahinkaya and Dilek (2007) point out that the inhibitory effect of some wastewater is attenuated when slow fillings are used with minimum durations of one hour, because they prevent the accumulation and the addition of high concentrations of the inhibiting substances, improving, in this way, the removal of contaminants in the SBRs.

Table 12. Efficiency of TKN removal in the SBRs for the treatments tested

Treatment	Total Kjeldahl Nitrogen (TKN)		Organic Nitrogen* (N_{org})	
	SBR Inlet ($mg \cdot L^{-1}$)	SBR outlet ($mg \cdot L^{-1}$)	Total removal (%)	N_{org}/NTK (%)
AI_L^1	162.8 ± 10.5	42.8 ± 7.0	73.7 a	92.6 a
CND_L^2	162.8 ± 10.5	48.7 ± 11.9	69.9 ab	90.3 ab
PD_L^3	141.8 ± 6.7	46.1 ± 1.2	67.5 abc	90.0 ab
NDS_R^4	164.5 ± 12.1	54.8 ± 9.6	66.8 abc	86.6 ab
PD_R^5	164.5 ± 12.1	56.0 ± 10.7	66.0 abc	89.9 ab
CND_E^6	127.6 ± 6.2	46.7 ± 2.0	60.0 bcd	69.5 d
AI_E^7	132.1 ± 12.6	48.4 ± 8.6	63.1 bcd	82.1 bc
AI_R^8	119.0 ± 12.8	45.0 ± 4.1	62.0 bcd	93.8 a
CND_R^9	119.0 ± 12.8	55.7 ± 4.6	55.1 def	92.8 a
NDS_L^{10}	141.8 ± 6.7	64.2 ± 8.0	54.7 def	68.5 d
PD_E^{11}	127.8 ± 3.5	61.3 ± 6.7	52.0 ef	71.4 d
NDS_E^{12}	127.8 ± 3.5	63.0 ± 1.9	50.1 f	74.2 cd

[1]AI_L: Intermittent aeration, slow filling, [2]CND_L: Conventional nitrification denitrification, slow filling, [3]PD_L: Pre-denitrification, slow filling, [4]NDS_R: Simultaneous nitrification denitrification, fast filling, [5]PD_R: Pre-denitrification, fast filling, [6]CND_E: Conventional nitrificationl denitrification, filling by stages, [7]AI_E: Intermittent aeration, stepwise filling, [8]AI_R: Intermittent aeration, fast filling, [9]CND_R: Conventional nitrification denitrification, fast filling,[10]NDS_L.Simultaneous nitrification denitrification, slow filling, [11]PD_E: Predesnitrification, stepwise filling, [12]NDS_E: Simultaneous nitrification denitrification, filling by stages.
*Mean followed by different letters in each column indicates significant differences according to the Tukey test (P≤0.05).

For the best treatments, the outlet concentration of TKN ranged between 42.8 and 56.0 $mg \cdot L^{-1}$. Particularly, in the case of the best treatment (AI_L), the TKN consisted of 40 $mg \cdot L^{-1}$ of organic nitrogen and only 2.8 $mg \cdot L^{-1}$ of ammoniacal nitrogen. For the rest of the cases, the behavior was similar, representing organic nitrogen between 86.6 and 93.8% of the NTK. These results showed that the SBR was able to nitrify practically all of the ammoniacal nitrogen present in the effluent (Table 12), however, it was not able to improve the ammonification of the important content of organic nitrogen that remained after the treatment in the SBR. This concentration probably corresponds to recalcitrant, or non-biodegradable nitrogen, that could not be hydrolyzed (van Lier et al., 2008).

Boursier et al. (2005) and Insel et al. (2009) noted that in industrial effluents, both from slaughterhouses and tanneries, organic nitrogen is in the particulate form, formed by complex organic polymers and probably not nitrifiable, severely limited the ammonification process. On the other hand, Kabdasli et al. (1993) reported the low transformation of organic nitrogen in tannery effluents, but observed that when they used manganese sulphate (II) to remove sulfides, hydrolysis of organic nitrogen was improved.

Comparing the efficiency removals of COD, TKN and $N-NH_4^+$ it was found that AI_L, PD_R, NDS_R and CND_L were treatments that removed the highest percentage of these three variables in a simultaneous way. However, it should be studied in greater detail in order to determine if the nitrified nitrogen (NO_2^-, NO_3^-) can be removed from the effluent by denitrification.

Denitrification efficiency was evaluated for each of the treatments and it was found that the highest percentage for the CND_R (Table 13), was statistically similar to NDCL ($P>0.05$), but higher than all the other treatments ($P\leq0.05$). In general, the most efficient sequence for dentrification was the CND (72.7%), followed by AI (60.3%) and PD (58.0), which showed no difference between them; while the least efficient was the NDS sequence (35.4%). Regarding the filling speed, the denitrification efficiency was higher when the filling was fast (70.6%), intermediate when slow filling was used (61.0%) and lower when the filling was used in stages (38.2%).

Denitrification efficiency is related to the content of $N-NO_2^-$ and $N-NO_3^-$ ($N-NO_x^-$) present at the end of the treatment, and is a function of the amount of ammoniacal nitrogen nitrified. In addition, it establishes the removal of $N-NO_x^-$ based on the $N-NO_x^-$ produced during the oxic phase. It is observed that the treatments that used rapid filling achieved greater denitrification efficiencies, coinciding with what was pointed out by Li et al. (2008), who maintained that in this way the consumption of CODFB is avoided in this first stage, favoring its use during the anoxic phase, and therefore, increasing the denitrification efficiency.

Table 13. Denitrification efficiency for the treatments studied

Treatment	N-NH$_4^+$ nitrified (mg·L^{-1})	N-NO$_x^-$ remaining (mg·L^{-1})	Denitrificación efficiency* (%)	Denitrificación rates* (mg N-NO$_3^-$ · mg SSV^{-1}·d^{-1})
CND$_R$[1]	61.9 ± 4.9	4.2 ± 1.4	93.2 a	0.0100 g
CND$_L$[2]	49.9 ± 11.8	6.8 ± 1.6	86.4 a	0.0370 cd
AI$_R$[3]	46.4 ± 3.8	10.0 ± 6.4	78.3 b	0.0104 g
AI$_L$[4]	50.9 ± 3.7	12.4 ± 3.4	75.6 b	0.1120 a
PD$_R$[5]	44.8 ± 1.5	15.8 ± 3.0	64.7 c	0.0140 fg
PD$_L$[6]	37.4 ± 1.8	15.0 ± 2.4	59.9 c	0.0190 ef
PD$_E$[7]	26.4 ± 4.7	13.3 ± 0.6	49.6 d	0.0420 c
NDS$_R$[8]	43.7 ± 7.7	23.8 ± 6.3	45.5 de	0.0130 g
NDS$_E$[9]	27.3 ± 4.0	16.4 ± 4.4	39.9 e	0.0030 h
CND$_E$[10]	38.4 ± 4.0	23.9 ± 4.8	37.8 e	0.0330 d
AI$_E$[11]	42.5 ± 4.0	31.9 ± 4.5	24.9 f	0.0845 b
NDS$_L$[12]	33.3 ± 4.8	26.6 ± 3.4	20.1 f	0.0220 e

[1]CND$_R$: Conventional nitrification denitrification, fast filling, [2]CND$_L$: Conventional nitrification denitrification, slow filling, [3]AI$_R$: Intermittent aeration, fast filling, [4]AI$_L$: Intermittent aeration, slow filling, [5]PD$_R$: Predesnitrification, fast filling, [6]PD$_L$: Predesnitrification, slow filling, [7]PD$_E$: Predesnitrification, stepwise filling, [8]NDS$_R$: Simultaneous nitrification denitrification, fast filling, [9]NDS$_E$: Simultaneous nitrification denitrification, stepwise filling, [10]CND$_E$: Conventional nitrification denitrification, stepwise filling, [11]AI$_E$: Intermittent aeration, stepwise filling, [12]NDS$_L$: Simultaneous nitrification denitrification, slow filling.
*Mean followed by different letters in each column indicates significant differences according to the Tukey test (P≤0.05).

The specific denitrification rates were variable depending on the treatment applied, most of them were higher than 0.00624 mg·mg^{-1}·d^{-1}, which was the specific rate reported by Ganesh et al. (2006) for effluents of tanneries. On the other hand, Farabegoli et al. (2004) reported specific denitrification rates of 0.422 mg·mg^{-1}·d^{-1} for a pretreated effluent that did not contain chromium, nor sulfides, being significantly higher than those obtained in the present investigation, showing that the presence of inhibitory substances could affect the speed of removal of N-NO$_x^-$. Finally, Ganesh et al. (2006) showed that to establish the exact denitrification rate, a complete nitrogen balance would be needed, including the measurement of the gaseous nitrogen formed.

An important observation was that denitrification occurred in all treatments to a greater or lesser extent, but without the need of addition of an external source of carbon, such as methanol or semi-treated waste water

(Metcalf and Eddy, 1995; Corbitt, 2003), which indicated that the organic matter required for denitrification was obtained by one of these sources:

- CODFB of the tannery wastewater for the NDS and PD sequences that were always initiated with an anoxic phase (Ganesh et al., 2006; Szpyrkowicz and Kaul, 2004).
- CODFB obtained after hydrolysis of the slowly biodegradable fraction of the COD, particularly in the treatments that used CND and AI sequences, which were initiated with oxic phases, and therefore, there was enough time for the CODLB to become CODFB (Orhon et al., 1999a; Ekama and Wentzel, 2008b).
- Internal carbon present in the tissue fragments of microorganisms that remain in the mixed liquor after endogenous degradation (Metcalf & Eddy, 1995; Garzón-Zúñiga, 2005). As well as the organic carbon accumulated inside microorganisms that can be used by denitrifiers during the anoxic phase (Obaja et al., 2005; Ciggin et al., 2007; Li et al., 2008).

The strategy of filling by stages was used, in order to supply easily biodegradable organic matter at the beginning of the anoxic phases, thus improving the denitrification process. However, this aeration sequence did not produce the expected results (Table 13). This behavior could be attributed to the fact that the limiting factor of the denitrification process was not the availability of biodegradable organic matter, but other factors such as the concentration of denitrifying microorganisms present in the biomass, concentration of NO_2^- and NO_3^-, available to denitrify dissolved oxygen concentrations, pH, and the presence of inhibiting compounds, among others (Metcalf and Eddy, 1995; Garzón-Zúñiga, 2005).

There are several ways to determine if the content of organic matter limits the denitrification process, one of which is the denitrification potential of the system. For industrial effluents, this potential is usually expressed in terms of the relationship between CODBT and the concentration of ammoniacal nitrogen in raw wastewater (Boursier et al., 2005).

Based on the above, for tannery effluents this potential ranged between 10.4 and 11.3 $g \cdot g^{-1}$, being higher than that recommended by Boursier et al. (2005), who achieved complete denitrification in slaughterhouse effluents when the ratio was greater than 5 $g \cdot g^{-1}$. On the other hand, Carucci et al. (1999) and Szpyrkowicz and Kaul (2004) established that, for tannery effluents, the COD/TKN ratio should be 8 and 7 $g \cdot g^{-1}$, respectively, at a minimum, to guarantee the wastewater will contain enough organic carbon during the denitrification stage. For this, in the present research it was found that the COD/TKN ratio was between 10.1 and 14.7, which shows that the presence of organic matter available for denitrification was not the factor that affected its performance.

When evaluating the cycles of each of the treatments, it was found that the concentration of dissolved oxygen (DO) ranged between 0.15 and 0.21 $mg \cdot L^{-1}$ for the treatments that started with anoxic phase (NDS and PD), while these varied between 0.17 and 0.69 $mg \cdot L^{-1}$ for those who started with the oxic phase. As expected, the anoxic conditions were more difficult to reach for those treatments with a previous oxic phase. However, the results obtained showed that the denitrification efficiency was not affected (Table 13), the best treatments being those that started with the oxic phase (CND and AI), in such a way that the concentration of dissolved oxygen apparently did not affect the process of denitrification.

In parallel, the behavior of the pH during the treatments was evaluated and it was observed that the pH of the mixed liquor was maintained between 8 and 8.9 units, this being adequate to guarantee that the total denitrification which occurs and the final product will be N_2 (pH> 7.3). Szpykowicz and Kaul (2004) pointed out that for effluents of tanneries, the variation of pH only affected, significantly, the nitrification process and had no effect on denitrification.

On the other hand, the denitrification could have been affected by the presence of inhibiting substances such as trivalent chromium, chlorides and sulfides in the effluents, whose concentrations were at 13.4 ± 8.6, $8.549.5 \pm 1,668.8$ and 147.5 ± 6.5 $mg \cdot L^{-1}$, respectively. However, it is important to note that Farabegoli et al. (2004) showed that the community of denitrifying microorganisms is less sensitive to the dimunition of their activity due to the presence of inhibitory substances, such as chromium, in comparison to the population of nitrifying microorganisms. On the other hand, Szpykowicz and Kaul (2004) observed that denitrification was not affected by the presence of high salinity of the effluent (≤ 5 g NaCl $\cdot L^{-1}$).

Despite this, the results showed that there was a good correspondence between the production of N-NOx- and the consumption of them for most of the treatments (Figure 10), and it was observed that CND_R, NDC_L and AI_R almost achieved total denitrification of all nitrified ammoinum, leaving a low residual of nitrogen product of the biological treatment in the effluent (4.2, 6.8 and 10 mg$\cdot L^{-1}$ N-NO$_x^-$, respectively). In particular, these treatments complied with the discharge limit established in the Venezuelan regulations that allows a maximum of 10 mg$\cdot L^{-1}$ of nitrogen in the form of N-NO$_2^-$ and N-NO$_3^-$. It was found that the less favored treatments were those that used filling by stages and those that used the NDS strategy.

Insel (2007), when performing the simulation of the NDS sequence, determined that the efficiency of denitrification, as well as that of nitrification, depends directly on the concentration of dissolved oxygen, as well as on the chosen aeration sequence. Therefore, the appropriate selection of the dissolved oxygen level together with an adequate control of the aeration strategy were the key in maximizing the efficiency of nitrification and denitrification. The low nitrogen removal efficiency obtained could be related to the selection of the concentration range of dissolved oxygen used in the oxic phase, which varied between 0.39 and 0.88 mg$\cdot L^{-1}$.

On the other hand, Third et al. (2003) reported that the denitrification efficiency drastically decreased when the ammonia nitrogen was completely consumed, attributing this behavior to the fact that, apparently the ammonium serves as a protective shield of the heterotrophic cells, and when it is consumed, the oxygen penetrates more deeply in the biomass flocs, eliminating the anoxic microzones that existed inside.

Based on the results obtained in the present research, it should be noted that the denitrification did not occur completely in any of the treatments, due to the combination of the conditions mentioned above, such as the filling form of the reactor (filling by stages was the least efficient strategy), pH, DO and the presence of inhibitory substances. It can be concluded that in the treatments in which simultaneous nitrification-denitrification (NDS) was favored, an adequate concentration gradient of dissolved oxygen was not achieved, which allowed regions with high concentrations of dissolved oxygen to exist in the biomass flocs, and other areas with active zones of low concentration of dissolved oxygen, so that denitrification could not occur to a large extent (Ganesh et al., 2006).

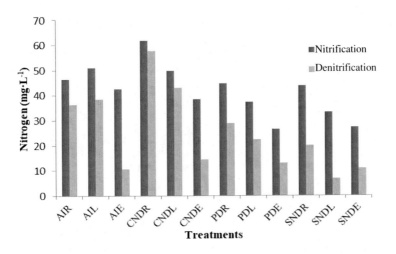

Figure 10. Comparison between the amount of nitrified and denitrified nitrogen for each treatment. AI: intermittent aeration, CND: nitrification and conventional denitrification, PD: pre-denitrification, SND: nitrification and simultaneous denitrification, R: fast filling of the SBR, L: slow filling of the SBR, E: stage filling of the SBR.

Efficiency of the SBR in the Removal of Total Nitrogen

Total nitrogen (TN) removal was favored by the PD_R treatment, achieving TN removal efficiencies of 60.6%, being statistically similar to those obtained with the CND_R, NDS_R, NDC_E treatments, whose removal was between 48.3 and 53.8% (Table 14). All these values were higher than the rest of the treatments (P≤0.05), with the lowest total nitrogen removal for the AI_R treatment (23.9%).

Table 14. Total nitrogen removal efficiency for the treatments studied

Treatment	Total nitrogen (TN)		
	SBR inlet (mg·L^{-1})	SBR outlet (mg·L^{-1})	Total removal* (%)
PD$_R$[1]	165.6 ± 12.7	65.2 ± 5.3	60.6 a
CND$_R$[2]	173.7 ± 43.6	79.3 ± 16.1	53.8 ab
NDS$_R$[3]	165.6 ± 12.7	85.7 ± 10.9	48.4 abc
CND$_E$[4]	132.1 ± 9.2	73.4 ± 3.5	48.3 abc
PD$_L$[5]	144.5 ± 6.6	78.8 ± 6.2	45.4 bcd
PD$_E$[6]	129.0 ± 3.6	72.5 ± 3.6	43.8 bcde
CND$_L$[7]	165.7 ± 9.4	99.4 ± 12.0	40.0 cde
AI$_L$[8]	165.7 ± 9.4	100.2 ± 4.6	39.4 cde
NDS$_E$[9]	129.0 ± 3.6	89.2 ± 12.3	35.8 def
NDS$_L$[10]	141.8 ± 6.6	97.0 ± 2.1	32.7 ef
AI$_E$[11]	132.4 ± 7.6	89.8 ± 10.7	32.2 ef
AI$_R$[12]	173.8 ± 43.6	132.2 ± 32.7	23.9 f

[1]PDR: Pre-nitrification, fast filling, [2]CND$_R$: Conventional nitrification denitrification, fast filling, [3]NDS$_R$: Simultaneous nitrification denitrification, fast filling, [4]NDC$_E$: Conventional nitrification denitrification, stepwise filling, [5]PD$_L$: Pre-denitrification, slow filling, [6]PD$_E$: Pre-denitrification, filling by stages, [7]CND$_L$: Conventional nitrification denitrification, slow filling, [8]AI$_L$: Intermittent aeration, slow filling, [9]NDS$_E$: Simultaneous nitrification denitrification, stepwise filling, [10]NDS$_L$: simultaneous nitrification denitrification, slow filling, [11]AI$_E$: Intermittent aeration, stepwise filling, [12]AI$_R$: Aeration intermittent, fast filling.
*Mean followed by different letters in each column indicates significant differences according to the Tukey test (P≤0.05).

Statistical analysis showed that the fast filling was significantly higher (46.6%) in the treatments that used slow fillings (39.4%) and by stages (40.0%), which did not show a statistical difference between them (P>0.05). Regarding the aeration sequence, it was found that the CND and PD strategies, which removed 47.4 and 49.9%, respectively (Table 13), were significantly higher in total nitrogen removal than the NDS strategy (39.0%) and AI strategy (31.8%).

In the treatments that used fast fillings TN removal was greater, independently of the sequence of aeration used, except for the intermittent aeration sequence, which indicated that for this particular treatment, the factor that controlled the removal of TN was the aeration sequence and not the filling time, as happened in the other cases.

Ganesh et al. (2005) indicated that the presence of biodegradable organic matter in the wastewater of the tanneries (66-70%) could be sufficient to achieve complete denitrification; however, their system only removed 36.3% of total nitrogen, so they suggested adding an additional filling, under anoxic conditions, to optimize denitrification. The present investigation included anoxic filling during the sequences of simultaneous nitrification-denitrification (NDS) and predesnitrification (PD), and a total nitrogen removal of between 48.4 and 60.6% was obtained when fast filling was used (Table 14), being greater than the 36.3% mentioned above.

On the other hand, the overall nitrogen balance for the treatments that used the intermittent aeration sequence had low efficiency in relation to that obtained by Insel et al. (2009), who also used AI for the treatment of tannery effluents and achieved 60% removal. The reason for this low efficiency may have been because of the minimum anoxic volume required in any intermittent aeration process was not achieved, and therefore, the denitrification efficiency was low. In addition to this, the remaining ammoniacal nitrogen, this could not be nitrified because the air time of the last sub-cycle was insufficient; it was also quantified in the nitrogen balance. All this contributed to the fact that, at the end of the cycle in the SBR, the total nitrogen content was higher than the rest of the treatments.

The treatment that used the conventional sequence for nitrogen removal (CND) was effective especially when rapid filling was used (Table 14). These results suggest that this sequence of aeration allowed the slowly biodegradable fraction of the COD of the raw wastewater to be hydrolysed, and to be available during the anoxic phase, favoring the removal of total nitrogen from the system. This behavior coincided with that reported by Ciggin et al. (2007), who pointed out that the fast and phased fillings favor the "abundance" and "lack" regimes. In these regimes the microorganisms rapidly remove the external substrate and produce storage polymers (period of abundance). When the external substrate is finished, the deficiency period begins, in which the microbial growth occurs at the expense of the accumulated carbon, denitrification taking place by using these polymers as a carbon source.

After the analysis was carried out, it should be noted that the TN removal efficiency was limited by the high organic nitrogen content remaining at the end of the treatments. Faced with this situation, none of the sequences and filling times used had an influence sufficient to achieve the process of ammonification of the organic nitrogen of the wastewater. In this regard, Insel et al. (2009), from a kinetic study, concluded that the process of ammonification is the limiting stage in the removal of nitrogen for the effluents. The low rate of ammonification is recognized as a specific characteristic of tannery effluents, due to the high content of organic nitrogen present in the proteins (Kabdasli et al., 1993; Orhon et al., 1999a; Insel et al., 2009). In addition, Insel et al. (2009) pointed out that the kinetic limitation produces a residual concentration of organic nitrogen that ranges between 12 and 25 mg·L^{-1}, being lower than that found in the present investigation (40-52 mg·L^{-1}), this being one of the main reasons why no greater removal of total nitrogen was obtained.

When considering the total nitrogen removal, as well as COD, the treatments that proved to be efficient were PD_R and NDS_R. Particularly, Li et al. (2008) argued that short filling times favor the removal of nitrogen, because the consumption of CODFB during the filling phase is avoided, and this fraction is reserved for use during denitrification in SBR, operating under the pre-denitrification strategy. In parallel, Durai and Rajasimman (2011) pointed out that a pre-denitrification process is more efficient for the simultaneous removal of nitrogen and organic matter in wastewater from tanneries. Based on this report, and considering that the efficiencies achieved for the removal of COD were not very different from each other, on average, most of the treatments were efficient enough to remove the biodegradable fraction of the COD, then the efficiency of nitrogen removal was used a decision criteria for the selection of the most optimal treatment.

Based on the above, the pre-denitrification sequence was selected as the best treatment using fast filling (5 min), which removed a slightly higher concentration of N-NOx- than NDS_R. Finally, the best operational conditions for SBR are PD_R, 12 h of cycle, 30 h of HRT and 15 d of sludge age.

PHYSICOCHEMICAL POST-TREATMENT FOR TANNERY EFFLUENT

In this section the effect of the application of a physicochemical treatment to the effluent of the tannery is evaluated, following the biological treatment in the SBR, with the objective of eliminating the non-biodegradable organic matter and the remaining nitrogen. The physicochemical treatment used was coagulation-flocculation, using ferric chloride combined with brine bitter.

Characterization of Coagulants

The physicochemical postreatment was carried out using a mixture of ferric chloride (coagulant) and brine bitter as an adjuvant. Cegarra (2011) reported that the combined use of these compounds in the physicochemical treatment of the tannery wastewater produced the best removal of COD and organic nitrogen remaining from the biological process in the SBR. The brine bitter is an agroindustrial waste that showed its efficiency in the physicochemical treatment, since at concentrations of 100 g·L^{-1} it was a rich source of magnesium ions (1,300 mg·L^{-1}). In this regard, Arboleda (1992) and Andia (2000) noted that magnesium ions contribute significantly in the coagulation-flocculation processes, because they follow the same mechanism as aluminum ions when used as a coagulant.

Ferric chloride is a chemical coagulant that has been widely used to treat wastewater (Song et al., 2004). Responsible for the coagulation efficiency are Fe^{3+}ions (Table 15), which are characterized by being ten times more effective than bivalent ions (Degremont, 1979; Arboleda, 1992). The combination of these factors permitted the explanation of the synergic effect of these two compounds.

The characterization of the coagulants reflected the high content of chlorides that can contribute to the effluent; however, considering that the wastewater contains chlorides in concentrations of 8,549.5 mg·L^{-1}, the increase that would represent the use of the chemical and natural coagulant is not significant in the doses used (50 mL·L^{-1} for FeCl$_3$ and 5 mL·L^{-1} for brine bitter).

Table 15. Characterization of the coagulants used

Variable	Units	Ferric chloride (1,62 g·L^{-1}) ($\overline{X} \pm SD$)	Bitter brine (100 g·L^{-1}) ($\overline{X} \pm SD$)
pH.	---	1.2 ± 0.1	4.9 ± 0.1
Color	U.C Pt-Co	3,000 ± 14.7	5.0 ± 0.5
Turbidity	N.T.U	14.5 ± 0.8	ND[1]
Calcium hardness	mg·L^{-1}	400 ± 18.1	200 ± 10.1
Magnesium hardness	mg·L^{-1}	600 ± 45.1	1,300 ± 230.1
Chlorides	mg·L^{-1}	7,622 ± 177.3	33,500 ± 177.3

Table adapted from Cegarra (2011). [1]ND: Not detectable.

Another important characteristic to highlight was the color of the coagulant product of the oxidation of the iron in the prepared solution, both the color and the turbidity were higher for the ferric chloride than for the brine bitter solution, which was colorless and without appreciable turbidity. Finally, an important difference was the pH between the coagulant and adjuvant; the ferric chloride presented very acidic character, while the brine bitter was slightly acidic. The effect of these last characteristics (Cl^{-}, color and pH) is diluted when they are added to the effluent to be treated; however, its determination was made in order to establish the basic composition of the substances.

Selection of pH Suitable for Physicochemical Treatment

An important factor in the efficiency of the coagulant during the coagulation-flocculation process is pH. For ferric chloride, several pH ranges have been reported ranging from 4 to 12, highlighting in particular the range between 4 and 6, and values higher than 8 units (Degremont, 1979; Arboleda, 1992), while the brine bitter was efficient at pH between

11 and 12 (Roman, 2010). Due to the diversity of pH reports, it was decided to determine the most suitable pH for post-treatment of the tannery's effluents, considering that the coagulant and adjuvant were added mixed.

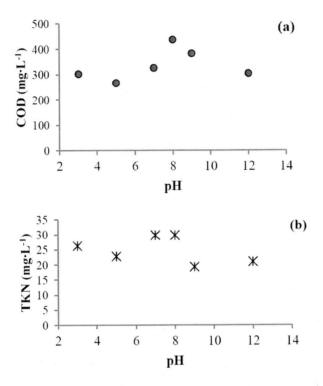

Figure 11. Behavior of the concentration of the chemical oxygen demand, COD (a) and total nitrogen Kjeldahl, TKN (b) after the physicochemical treatment evaluated at different pH.

According to the previously mentioned, there is a wide range of pH, consequently the coagulation-flocculation test was carried out at different pH, and the effect on the removal of COD and TKN was determined. The results showed that at pH 5 the highest COD removal was reached (60.4%), followed by 54.9% achieved at pH between 11 and 12 units, while the treatments at pH 8 and 9 were the least efficient, with removals of only 35 and 43%, respectively (Figure 11a). This indicated that the proper pH to remove COD in the post-treatment is in a wide range from 3 to 7 and 11 to 12 pH units. However, the objective of physicochemical treatment not only includes the removal of COD, but the joint decrease of organic matter and nitrogen remaining from the biological treatment process.

Regarding the removal of the organic nitrogen remaining from the effluent after the biological treatment, the TKN results can be observed at the end of the physicochemical treatment using different pH (Figure 11b). It was found that the highest removal efficiencies were achieved for alkaline pH (9 to 12 units) with removals of 50 and 45.5%, respectively. The lowest efficiency was obtained for pH close to neutrality, with only 22.7% removal, while the lower pH evaluated (3 and 5) showed intermediate removals of TKN (31.8 and 40.9%, respectively).

According to the results obtained for the concentrations of COD and TKN at the end of the physicochemical treatment, it was decided that the optimum pH range to use for the mixture of ferric chloride with brine bitter was between 11 and 12 units, in order to obtain greater removal efficiency, jointly. This behavior coincided with that reported by other researchers who reported the best performance of physicochemical treatment when coagulation-flocculation was performed at alkaline pH (Song et al., 2004; Ryu et al., 2007; Román, 2010; Hernández et al., 2016).

Selecting the Dose of Coagulants

The dose of the coagulants used in the research was based on the results obtained by Cegarra (2011) who carried out a factorial study which varied the doses and the concentration of three different coagulants (ferric chloride, brine bitter and seawater). The best treatment for COD removal was represented by the ferric chloride-bitter brine mixture with removal of 62.8%, while the seawater-ferric chloride mixture was more efficient for removing TKN (47.1%). Of these two treatments, Cegarra (2011) selected the mixture of 50 mL of ferric chloride (1.62 mg·L^{-1}) with 5 mL of brine bitter (100 g·L^{-1}), because the removal efficiencies were adequate to meet the regulations for the discharge to water bodies, and additionally because the final volume of coagulant and adjuvant was lower than when seawater was used (100 mL·L^{-1}) and ferric chloride (25 mL·L^{-1}).

According to the results obtained by Cegarra (2011), the averages of COD and TKN after physicochemical treatment were 334.6 and 30.8 mg·L^{-1}, respectively. Both results were very close to the limit established in the Venezuelan legislation, which establishes the maximum discharge limit to water bodies of 350 mg COD·L^{-1} and 40 mgTN·L^{-1}. Therefore, in the present investigation, a new dose of coagulant was tested to optimize the previous results. The volume of the chemical coagulant was slightly increased to 60 mL·L^{-1} in order to evaluate the effect in the removal of COD and TKN.

It was observed that the COD showed no variation in its final concentration (332.6 mg·L^{-1}). However, for the TKN the effect was important, generating an effluent with concentrations of 23.4 mg·L^{-1} (Table 16). Also, almost total removal of N-NH$_4^+$ were achieved (0.5 mg·L^{-1} remaining) and it was observed a small decree in the concentration of N-NO$_x^-$ from 17.2 to 11.7 mg·L^{-1}. The importance of the adjustment made to the coagulant dose was based mainly on the fact that the research conducted by Cegarra (2011) did not quantify the concentration of N-NO$_x^-$ because it was assumed that they had been completely transformed to N$_2$ during the biological treatment.

Table 16. Optimization of the coagulant dose (FeCl$_3$ 1.62 g·L^{-1}) maintain the adjuvant dose constant (5 mL bitter brine, 100 g·L^{-1})

Variable	60 mL de FeCl$_3$ (1,62 g·L^{-1})			50 mL de FeCl$_3$ (1,62 g·L^{-1})[1]		
	Physicochemical treatment (mg·L^{-1})		Removal (%)	Physicochemical treatment (mg·L^{-1})		Removal[1] (%)
	Before	After		Before*	After*	
COD	703.9 ± 94.0	332.6 ± 22.1	52.7 ± 5.6	911.5 ± 150.4	334.6 ± 46.5	62.8 ± 2.3
TKN	62.1 ± 15.4	23.4 ± 6.5	62.4 ±5.7	52.0 ± 3.3	30.8 ± 8.7	41.9 ± 3.6
N-NH$_4^+$	5.3 ± 4.4	0.5 ± 0.8	90.0 ± 9.3	NR[2]	NR[2]	---
N-NO$_x^-$	17.2 ± 6.9	11.7 ± 2.3	32.0 ± 6.9	NR[2]	NR[2]	---

[1]Results reported by Cegarra (2011), [2]NR. Concentrations not reported.

However, in this investigation, this was not achieved in its entirety, so the decrease in the concentration of nitrogen species was essential to comply with the legal regulations for discharge to water bodies (MARNR, 1995). In parallel, physicochemical variables such as pH, color, turbidity, chlorides, calcium hardness and magnesium hardness were measured (Table 17). It was observed that the optimized coagulant mixture reduced the turbidity and color of the effluent by 51.4 and 64.9%, respectively, improving the appearance of the wastewater after treatment (Figure 12). However, these variables have not yet met the quality limits established in the Venezuelan environmental regulations (MARNR, 1995).

Table 17. Behavior of physicochemical variables during treatment

Variable	Post-treatment		Removal (%)	
	Before $\bar{X} \pm SD$	After $\bar{X} \pm SD$	This research	Cegarra (2011)
pH.	8.96 ± 0.23	10.7 ± 1.8	---	---
Real color (U.C.Pt-Co)	2,366.8 ± 1.364.8	1,000.0 ± 707.1	51.4	40.0
Turbidity (N.T.U)	234.1 ± 99.9	154.4 ± 101.1	64.9	35.5
Calcium hardness (mg·L^{-1})	562.5 ± 467.9	300.0 ± 45.1	46.7	56.1
Magnesium hardness (mg·L^{-1})	1,337.5 ± 878.8	200.0 ± 141.4	81.3	86.8
Chlorides (mg·L^{-1})	8,696.7 ± 4.300.7	9,885.5 ± 4.616.8	13.7[1]	3.0[1]

[1]Correspond to the increase in the concentration by contribution of the coagulants.

On the other hand, it was observed that the calcium and magnesium hardness decreased at the end of the treatment, which indicates that the magnesium and calcium ions from the brine bitter contributed to the coagulation-flocculation process. Finally, it was observed that the new dose of ferric chloride in the coagulant mixture, improved the removal of color and turbidity, compared to the results obtained by Cegarra (2011), while the removal of calcium and magnesium hardness remained constant, due to the fact that the dose of bitter brine was not modified.

Figure 12. Physical appearance of the tannery wastewater. From left to right: raw wastewater, SBR effluent, effluent physical-chemical process.

It is important to note that concentration of chlorides increased by 13.7% after the addition of the coagulants; however, it is considered a non-significant increase, in comparison with the high content of these ions present in the raw tannery effluent, generated by the chemical reagents used during the production process of the leather.

Ryu et al. (2007) reported the efficiency of the physicochemical treatment using seawater as a coagulant to treat tannery effluents. They observed that a content of 1,200 mg $Mg^{+2} \cdot L^{-1}$ in seawater permitted COD and TKN removals of 67 and 43%, respectively, when a dose of 100 mL$\cdot$ L^{-1} at a pH of 11 was used in the coagulation-flocculation process. For the present investigation, the natural source rich in magnesium was the bitter brine adjuvant. It was found lower removals of COD, but higher removals of NTK in comparison with results obtained by Ryu et al. (2007).

Other researchers, such as Dosta et al. (2007) integrated post-treatment to an SBR, in which treated supernatant of anaerobically digested piggery wastewater was used. The process consisted of adding ferric chloride, as a coagulant, in a concentration of 0.8 $g \cdot L^{-1}$ at the end of the anoxic phase prior to sedimentation. The COD removal obtained was higher than 66%, but no variation in the nitrogen concentration was observed. This behavior could be attributed to the fact that the nature of its components was not colloidal. In the present investigation it was observed that the coagulation-flocculation process removed TKN, $N-NH_4^+$, and to a lesser extent $N-NOx^-$, due to the coagulation of the colloidal fraction that constituted each species of nitrogen mentioned. The tannery effluents are characterized by having AZO compounds, used for dyeing the skins, which can be attacked by the coagulants, causing a decrease of the total nitrogen at the end of the treatment.

Most of the literature consulted for the treatment of tanney wastewaters, indicate the use of physicochemical processes as pretreatments to remove chromium, sulfides, total solids, suspended solids, COD, among others, and to subsequently submit this wastewater to biological treatment (Orhon et al., 1999a; Graterol, 2000; Farabegoli et al., 2004; Ryu et al., 2007; Lefebvre et al., 2006). However, the results obtained in the present study, using coagulation-flocculation as post-treatment, optimized the amount of necessary coagulant, because it only needs to act on non-biodegradable content of COD and nitrogen, which were the major contaminants remaining from the biological process. In this way, the amount of chemical reagents required to produce an effluent that met the environmental standards established in Venezuelan legislation could be improved.

Finally, the sequence of biological-physicochemical treatments permitted for the maximum performance of each process, since the biological treatment eliminated the major part of the biodegradable COD fraction, while the physicochemical treatment removed the remaining (no biodegradable) material. It is also important to note that the treatment sequence worked because it was combined with a primary sedimentation treatment that occurred in the storage pond. In this first stage, the high trivalent chromium content generated in the industry was removed by natural chemical precipitation, as well as uncontrolled chemical and biological reactions that permitted the wastewater to be treated biologically in the SBR.

PROPOSAL OF A TREATMENT TRAIN FOR TANNERY WASTEWATER AT PILOT SCALE

The proposed scheme for the treatment of the wastewater of the tannery is presented in Figure 13. The system is composed of a primary, a secondary (biological) and a tertiary (physicochemical) treatment. The approach includes the use of the two ponds that currently exist in the tannery, one to store the raw effluent (pond 1) and the other to contain the treated wastewater (pond 2).

The proposal calls for the primary treatment to be carried out in pond 1. In this first stage of the treatment, it was found that chemical and biological reactions occurred that allowed for the reduction of the concentration of trivalent chromium, sulfides, COD and nitrogen. Therefore, it is recommended that in this stage, the process of natural sedimentation by gravity be maintained, as well as the mixing and equalization of the components of the effluent, which will later be transferred to the biological treatment unit.

The secondary treatment will be based on a sequential batch reactor that operates with suspended biomass, using the anoxic-oxic-anoxic aeration sequence (pre-denitrification) and fast filling of the reactor. These operational strategies were found to be the best during the experimentation phase of this research. The calculation of SBR pre-sizing was made for a total capacity of 3,000 L and a useful work volume is 2,000 L (Figure 14). The SBR mix liquor will be made up of 30% biomass (600 L) and 70% of wastewater (1,400 L). The duration of the cycle will be 12 h with a reactor volume exchange (VER) of 40%, corresponding to the best conditions found in this investigation. This means that 1,600 L of wastewater would be processed daily (800 L in each SBR cycle), it is estimated SBR will work with VOC of 1.2 $kg \cdot m^{-3} \cdot d^{-1}$ considering a tCOD of 1,500 $mg \cdot L^{-1}$.

The tannery produces an average of 6,000 L of raw wastewater per day. The SBR would process 1,600 $L \cdot d^{-1}$, and therefore 4,400 $L \cdot d^{-1}$ would be accumulated in the first pond. The maximum working capacity of pond 1 is 1,373 m^3, so after approximately 312 days it would reach its full capacity. However, the evaporation of the liquid should be considered, which would imply that the saturation time of the pond would increase.

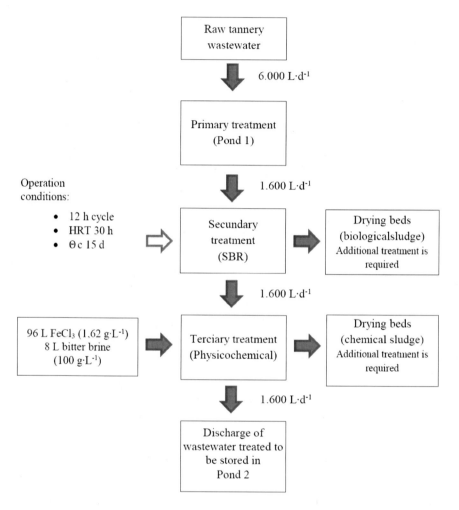

Figure 13. Proposal of treatment sequences for tannery effluent studied.

The other operating conditions, such as the hydraulic residence time in the SBR, would be 30 h, while the cellular retention time (sludge age) would be maintained at approximately 15 days. The importance of using this last time is because high sludge ages allow the enrichment of the biomass of nitrifying autotrophic microorganisms, which are characterized by having a slower growth rate than heterotrophs, thus favoring the realization of the biological processes for the removal of nitrogen (Ekama and Wentzel, 2008a; Carrasquero, 2011).

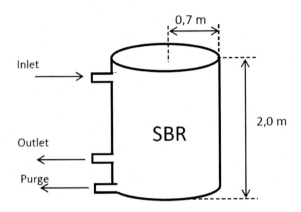

Figure 14. Dimensions of the sequential batch reactor. Pilot scale.

To maintain the age of the sludge at 15 d, a mixture liquor extraction process (purge) must be carried out at the end of the second anoxic phase. The volume to be purged would be 134 L per cycle of operation and would be accumulated in drying beds located on one side of the SBR. It is important to highlight that this sludge must be subjected to an adequate treatment before being finally disposed of. Figure 15 shows the stages and phases required for the proper functioning of the SBR.

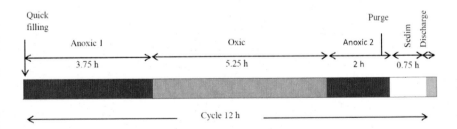

Figure 15. Stages of the SBR and the sequence of aeration during the reaction time.

The SBR will work automatically by means of timers that activate the charge, discharge and purge pumps, as well as the compressor for air supply and the motor for the agitation of the system. The SBR must be agitated to keep the biomass suspended during the reaction time and to achieve the best contact between the microorganisms and the liquid. The use of a frequency inverter coupled to an electric motor, connected to a blade submerged in the mixed liquor, is recommended (Stenstrom and Rosso, 2008).

On the other hand, it is necessary to supply air to the system, through the use of a compressor, during the oxic phase of the reaction time, in order to maintain the DO concentration above 2 $mg \cdot L^{-1}$ (Figure 15). The aeration must be done from the bottom of the reactor, using a network of perforated pipes with small holes that produce fine bubbles that allow a greater diffusion of the air during its ascent in the column of liquid. One of the main drawbacks of this type of aerator is that it requires periodic maintenance to prevent obstructions in the orifices (Stenstrom and Rosso, 2008).

The biological system can be controlled in real time, by means of pH, ORP and DO measuring probes. For this, the system must be automated so when the nitrification ends (detection of the ammonium valley in the pH profile), the air supply is suppressed and the next phase of the treatment is continued. In the same way, the automation must allow stoppage of the anoxic phase of the treatment, when in the profile of ORP and pH inflections are detected, that indicating that the denitrification process has finished. This control would allow for adjustment of the duration of each phase to the real needs of the treatment of the effluent.

Subsequently, after a cycle of biological treatment has been completed, 800 L of the treated effluent would be transferred to an inverted conical tank (hopper type), that will have an engine with a blade connected to a frequency inverter, similar to that used during the treatment in the SBR, where the coagulation-flocculation process would take place. In this unit the mixture of coagulants would be added and the pH would be adjusted between 11 and 12 units using 6 N NaOH. To carry out the post-treatment to the 800 L of wastewater, 48 L of $FeCl_3$ (1.62 $g \cdot L^{-1}$) is required and 4 L of brine bitter (100 $g \cdot L^{-1}$), corresponding to 129.6 g of $FeCl_3.6H_2O$ (common presentation of the salt) and 400 g of brine bitter. Coagulation will occur during the rapid mixing period and flocculation during slow mixing. The frequency converter will allow changes of the rotational speed of the motor, and finally stop the agitation of the sedimentation.

Finally, the clarified effluent from the post-treatment will be transferred to the second pond where it will remain stored. The sediments and precipitated material of the physicochemical process will be removed from the bottom of the tank placed in an area adjacent to the treatment plant, permitting adequate treatment. In this way the tank would be ready to receive the next load of wastewater.

REFERENCES

Abu-Ghararah, Z. H., Randall, C. W. (1990). Effect of influent organic compounds on biological phosphorus removal. *Journal of Water Science & Technology*, 23(4/6):585-594.

Ahn, D. H.; Chang, W. S., Yoon T. I. (1999). Dyestuff wastewater treatment using chemical oxidation, physical adsorption and fixed bed biofilm process. *Process Biochemistry*, 34(5):429-439.

Andía, Y. (2000). *Tratamiento de agua: Coagulación y floculación. Evaluación de Plantas y Desarrollo Tecnológico.* SEDEPAL. 44 p. [*Water Treatment: Coagulation and floculation. Assessment of plants and technological development.* SEDEPAL. 44 p].

Arboleda, V. (1992). *Teoría y Práctica de la Purificación del Agua: Teoría de la coagulación del agua.* Editorial ACODAL. Colombia. [*Theoric and Practice of Water Purification: Theory of water coagulation.* Editorial ACODAL. Colombia].

Arslan, A., Ayberk, S. (2003). Characterisation and biological tratability of "Izmit industrial and domestic wastewater treatment plant" wastewaters. *Water SA*, 29(4):451-456.

Barajas, M. (2002). *Eliminación biológica de nutrientes en un reactor biológico secuencial.* Trabajo Doctoral de Ciencias Biológicas. Universidad Técnica de Cataluña. Barcelona. 345 p. España. [*Biological nutrient removals in a sequential batch reactor.* Thesis in

Biological Sciences. Technical University of Cataluña. Barcelona. Spain. 345 p.].

Boursier, H., Béline, F., Paul E. (2005). Piggery wastewater characterization for biological nitrogen removal process design. *Bioresource Technology*, 96(3):351-358.

Cado O. (1996). Sales de cromo: Su relación con el medio ambiente. *Gerencia Ambiental* 3(30):770-778. [Chromium salts: Its relation with environment. *Environmental Management* 3(30):770-778].

Carrasquero, S. (2011). *Influencia del tiempo de retención celular y de la secuencia de fases en la etapa de reacción sobre la remoción de nutrientes en un SBR*. Trabajo de Grado. Posgrado de Ingeniería. La Universidad del Zulia. Venezuela. [*Influence of sludge age and phase sequencies during reaction stage on the nutrient removals in a SBR*. Master Program in Environmental Science. Zulia University. Venezuela].

Caruccí, A., Chiavola, A., Majone, M., Rolle, E. (1999). Treatment of tannery wastewater in a sequencing batch reactor. *Water Science and Technology*, 40(1):253–259.

Cegarra, D. (2011). *Tratamiento físico-químico en efluentes de una tenería provenientes de un tratamiento biológico*. Trabajo de grado. Maestría en Ciencias del Ambiente, mención Ingeniería Ambiental. Universidad del Zulia. Venezuela. 71 p. [*Physicochemical tannery wastewater as a posttreatment of a biological process*. Master Program in Environmental Science. Zulia University. Venezuela. 71 p.].

Ciggin, A., Karahan, Ö., Orhon, D. (2007). Effect of feeding pattern on biochemical storage by activated sludge under anoxic conditions. *Water Research*, 41:924-934.

Cokgör, E., Sözen, S., Orhon, D., Henze, M. (1998). Respirometric analysis of activated sludge behaviour-I. Assessment of the readily biodegradable substrate. *Water Research*, 32(2):461-475.

Corbitt, R. (2003). *Manual de Referencia de la Ingeniería Medioambiental*. Editorial McGraw Hill. 1608 p. España. [*Reference*

Manual of Environmental Engineering. Editorial McGraw Hill. 1608 p. Spain].

Degremont, G. (1979). *Manual Técnico del Agua.* Cuarta Edición. Editorial Urmo, S.A. de Ediciones. España. [*Water Technological Manual.* Fourth edition. Editorial Urmo, S.A. of Spain editions].

Dosta, J., Gali, A., Benabdallah El-Hadj, T., Mace, S., Mata-Alvarez, J. (2007). Operation and model description of a sequencing batch reactor treating reject water for biological nitrogen removal via nitrite. *Bioresource Technology*, 98:2065-2075.

Dosta, J., Rovira, J., Galí, A., Macé, S., Mata-Álvarez, J. (2008). Integration of a Coagulation/Floculation step in a biological sequencing batch reactor for COD and nitrogen removal of supernatant of anaerobically digested piggery wastewater. *Biosource Technology*, 99 (13):5722-5730.

Durai, G., Rajasimman, M. (2011). Biological Treatment of tannery wastewater – A review. *Journal of Environmental Science and Technology*, 4(1):1-17.

Ekama, G., Wentzel, M. (2008a). Nitrogen Removal. In: *Biological Wastewater Treatment: Principles, Modelling and Design.* Edited by M. Henze, M. C. M van Loosdrecht, G. A. Ekama and D. Brdjanovic. Published by IWA Publishing, London, UK.

Ekama, G., Wentzel, M. (2008b). Organic Material Removal. In: *Biological Wastewater Treatment: Principles, Modelling and Design.* Edited by M. Henze, M. C. M. van Loosdrecht, G. A. Ekama and D. Brdjanovic. Published by IWA Publishing, London, UK. 511 p. Capítulo 4. pp 53-86.

Galisteo, M., Hermida-Veret, S., Viñas-Sendic, M. (2002). *Tratabilidad aeróbica de efluentes de curtiembre en la etapa de terminación.* Montevideo, Uruguay. 7 p. [*Aerobic tratability of tannery wastewater from dyeing stage. Montevideo, Uruguay. 7 p.*].

Ganesh, R., Balaji, G., Ramanujam, A. (2006). Biodegradation of tannery wastewater using sequencing batch reactor – Respirometric assessment. *Bioresource Technology*, 97(15):1815-1821.

Garzón-Zúñiga, M. (2005). *Mecanismos no convencionales de transformación y remoción del nitrógeno en sistemas de tratamiento de aguas residuales.* Ingeniería Hidraúlica en México, 20(4):137-149. [*Non conventional mechanisms for transformation and removal of nitrogen in wastewater treatments.* Hydraulic Engineering in Mexico, 20(4):137-149].

Germili, F., Orhon, D., Artan, N. (1991). Assessment of the initial inert soluble COD in industrial wastewaters. *Water Science and Technology*, 23(4/6):1077-1086.

González, M., Saldarriaga, J. (2008). Remoción biológica de materia orgánica, nitrógeno y fósforo en un sistema tipo anaerobio. *Revista EIA*, 10:45-53. [Biological removal of organic matter, nitrógen and phosphorus in an anaerobic system. *EIA Journal*, 10:45-53].

Graterol, N. (2000). *Remoción de materia orgánica en un reactor biológico discontinuo secuencial operando con aguas residuales de una tenería.* Barquisimeto, Venezuela. 10 p. [*Organic matter removal in sequential batch reactor treating tannery wastewater.* Barquisimeto. Venezuela. 10 p].

Grunditz, C., Dalhammar, G. (2001) Development of nitrification inhibition assays using pure cultures of Nitrosomonas and Nitrobacter. *Water Research*, 35:433–440.

Gutierrez, J., Espino, A., Coreño, A., Acevedo, F., Reyna, G., Fernández, F., Tomasini, A., Wrobel, K., Wrobel, K. (2010). Mecanismos de interacción con cromo y aplicaciones biotecnológicas. Revista Latinoamericana de Biotecnología Ambiental Algal, 1(1):47-63. [Interaction mechanism with chromium and its biotechnological applications. *Latin-American Algal Environmental Biotechnologic Journal*, 1(1):47-63].

Hermida-Veret, S., Galisteo, M., Viñas, M. (2000). Evaluación respirométrica de la biodegradabilidad aeróbica de un efluente de curtiembre. *XXVII Congresso Interamericano de Engenharia Sanitária e Ambiental.* Trabajo en extenso I-184. pp 1-7. Porto Alegre. Brasil. [Respirometric evaluation of aerobic biodegradability of tannery efluente. *27th Interamerican Congress of Sanitary and Environmental Engineering*. I-184. Pp 1-7. Porto Alegre. Brazil].

INESCOP. Centro de Innovación y Tecnología (2008). *Aplicación de bioensayos respirométricos en aguas de tenerías*. Departamento del Medio Ambiente de INESCOP. Alicante, España. 16 p. [Innovation and Technological Center (2008). *Respirometric bioassays in tannery wastewater*. INESCOP Environmental Department. Alicante. Spain. 16 p.].

Insel, G., Sin, G., Lee, D., Nopens, I., Vanrolleghem, P. (2006). A calibration methodology and model-based systems analysis for SBRs removing nutrients under limited aeration conditions. *Journal of Chemical Technology and Biotechnology*, 81:679-687.

Insel, G. (2007). Effects of design and aeration control parameters on simultaneous nitrification and denitrification (SNdN) performance for activated sludge process. *Environmental Engineering Science*, 24(5):675-686.

Insel, H., Görgün, E., Artan, N., Orhon, D. (2009). Model based optimization of nitrogen removal in a full scale activated sludge plant. *Environmental Engineering Science*, 26(3):471-479.

Kabdasli, I., Tünay, O., Orhon, D. (1993). The treatability of chromium tannery wastes. *Water Science and Technology*, 28(2):97-105.

Kanagaraj, J., Chandra, N. K., Mandal, A. B. (2008). Recovery and reuse of chromium from chrome tanning wastewater aiming towards zero discharge of pollution. *Journal of Cleaner Production*, 16:1807-1813.

Karahan, Ö., Dogruel, S., Dulekgurgen, E., Orhon, D. (2008). COD fractionation of tannery wastewaters – Particle size distribution, biodegradability and modeling. *Water Research*, 42:1083-1092.

Kristensen, G., Jorgense, P., Henze, M. (1992). Characterization of functional groups and substrate in activated sludge and wastewater by AUR, NUR and OUR. *Water Science and Technology*, 25(6):43-57.

Lefebvre, O., Vasudevan, N., Torrijos, M., Thanasekaran, K., Moletta, R. (2005). Halophilic biological treatment of tannery soak liquor in a sequencing batch reactor. *Water Research*, 39(8):1471–1480.

Lefebvre, O., Vasudevan, N., Torrijos, M., Thanasekaran, K., Moletta, T. (2006). Anaerobic digestion of tannery soak liquor with an aerobic post-treatment. *Water Research*, 40(7):1492-1500.

Li, B., Irvin, S. (2007). The comparison of alkalinity and ORP as indicators for nitrification and denitrification in a sequencing batch reactor (SBR). *Biochemical Engineering Journal*, 34:248–255.

Li, J., Healy, M., Zhan, X., Rodgers, M. (2008). Nutrient removal from slaughterhouse wastewater in an intermittently aerated sequencing batch reactor. *Bioresource Technology*, 99(16):7644-7650.

Loayza, P. J. (2006). Remoción de iones metálicos por precipitación química. Boletín Electrónico Informativo sobre Productos y Residuos Químicos, 2(13):1-4. [Remotion of metalic ions for chemical precipitation. *Electronic Bulletin about Chemical Products and Residues*, 2(13):1-4].

MARNR. Ministerio del Ambiente de Recuersos Naturales Renovables. (1995). *Normas para la clasificación y el control de la calidad de los cuerpos de agua y vertidos o efluentes líquidos.* Gaceta oficial extraordinaria: 5.021 del 18/12/95. Decreto N° 883. Venezuela. [Ministry of Environment of Natural Renewal Resources. (1995). *Rules of classification and quality control of water bodies and wastewater.* 5,021 of December, 18, 1995].

Mekonnen, A., Leta, S. (2011). Effects of cycle and fill period length on the performance of a single sequencing batch reactor in the treatment of composite tannery wastewater. *Nature and Science*, 9(10):1-8.

Metcalf y Eddy. (1995). *Ingeniería de Aguas Residuales. Tratamiento, vertido y reutilización.* Volumen I y II. Tercera edición. Mc. Graw Hill/Interamericana de España, S. A. U. Madrid, España. [*Wastewater engineering. Treatment and reutilization.* Volumes I and II. Third edition. McGraw Hill. Madrid. Spain].

Munz, G., Gori, R., Cammilli, L., Lubello, C. (2008). Characterization of tannery wastewater and biomass in a membrane bioreactor using respirometric analysis. *Bioresource Technology*, 99:8612-8618.

Nemeron, N. (1977). *Aguas Residuales Industriales. Teorías, Aplicaciones y Tratamientos.* Primera edición. H. Blume Ediciones. Madrid,

España. [*Industrial Wastewaters. Theory, applications and treatments.* First edition. H. Blume Editions. Madrid. Spain].

Obaja, D., Mace, S., Mata-Alvarez, J. (2005). Biological nutrient removal by a sequencing batch reactor (SBR) using an internal carbon source in digested piggery wastewater. *Bioresource Technology*, 96: 7-14.

Orhon, D., Artan, N., Cimsit, Y. (1989). The concept of soluble residual product formation in the modeling of activated sludge. *Water Science and Technology*, 21:339-350.

Orhon, D., Artan, N., Ates, E. (1994). A description of three methods for the determination of the initial inert particulate chemical oxygen demand of wastewater. *Journal of Chemical Technology and Biotechnology*, 61(1):73-80.

Orhon, D., Genceli, E., Cokgör, E. (1999a). Characterization and modelling of activated sludge for tannery wastewater. *Water Environment Research*, 71(1):50-63.

Orhon, D., Karahan, O., Sözen, S. (1999b). The effect of residual microbial products on the experimental assessment of the particulate inert COD in wastewaters. *Waters Research*, 30(14):3191-3203.

Orhon, D., Sözen S., Cokgör, E. (1999c). Experimental basis for the hydrolysis of slowly biodegradable substrate in different wastewaters. *Water Science and Technology*, 39(1):87–95.

Orhon, D., Germili, F., Karahan, O. (2009) *Industrial wastewater treatment by activated sludge.* IWA Publishing. Londres. Reino Unido. 113-126.

Palma, M., Manga, J. (2005). Simulación de un sistema de fangos activados en discontinuo (SBR) para el tratamiento de aguas residuales con altos contenidos de nitrógeno. *Ingeniería y desarrollo*, 2(18):61-71. [Simulation of activated sludge in a sequential batch reactor for treatment of wastewater rich in nitrogen. *Engineering and Developments*, 2(18):61-71].

Palmero, J., Pire, M. C., Hernández, J. L., López, F. L., Rincón, N. C., Díaz, A. R. (2009). Fraccionamiento de la materia orgánica de un agua residual de la industria avícola para la remoción biológica de nutrientes. *Boletin del Centro de Investigaciones Biológicas*,

43(2):211-224. [Fractionation of organic matter of poultry wastewater for biological removal of nutrients. *Biological Research Center Bulletin*, 43(2):211-224].

Park, J., Wang, J., Novotny, G. (1997). Wastewater characterization for evaluation of biological phosphorus removal. Wisconsin Department of Natural Resources. *Research Report*. 174. 29 pp.

Pire-Sierra, M. C., Palmero, J., Araujo, I., Díaz, A. (2010a). Tratabilidad del efluente de una tenería usando un reactor por carga secuencial. *Revista Científica*, FCV-LUZ, XX(3):284-292. [Treatability of tannery effluent using sequential batch reactor. *Scientific Journal*, FCV-LUZ, XX(3):284-292].

Pire-Sierra, M. C., Palmero, J., Araujo, I., Díaz, A. (2010b). Tratabilidad del efluente de una tenería con presencia de cromo usando un reactor por carga secuencial. *Revista Científica*, FCV-LUZ, 20(4):390-398. [Treatability of tannery effluent rich in chromium using sequential batch reactor. *Scientific Journal*, FCV-LUZ, 20(4): 390-398].

Pire-Sierra, M. C., Rodríguez, K., Fuenmayor, M., Fuenmayor, Y., Acevedo, H., Carrasquero-Ferrer, S. J., Díaz, A. (2011). Biodegradabilidad de las diferentes fracciones de agua residual producidas en una tenería. *Ciencia e Ingeniería Neogranadina*, 21(2):5-19. [Biodegradability of different fractions of tannery wastewater. *Neogranadine Science and Engineering*, 21(2):5-19].

Ryu, H, Lee, S, Chung, K. (2007). Chemical Oxygen demand removal efficiency of biological treatment process treating tannery wastewater following seawater flocculation. *Environmental Engineering Science*, 24(3):394-399.

Sahinkaya E., Dilek, F. (2007). Effect of feeding time on the performance of a sequencing batch reactor treating a mixture of 4-CP and 2,4-DCP. *Journal of Environmental Management*, 83:427-436.

Song, Z., Williams, C., Edyvean, R. (2004). Treatment of tannery wastewater by chemical coagulation. *Desalination*, 164:249-259.

Stenstrom, M. K., Rosso, D. (2008). Aeration and Mixing. In: *Biological Wastewater Treatment: Principles, Modelling and Design*. Edited by

M. Henze, M. C. M. van Loosdrecht, G. A. Ekama and D. Brdjanovic. Published by IWA Publishing, London, UK.511 p. Chapter 9.

Stoop, M. L. M. (2003). Water management of production systems optimised by environmentally oriented integral chain management: case study of leather manufacturing in developing countries. *Technovation*, 23, 265-278.

Szpyrkowicz, L., Kaul, S. (2004). Biochemical removal of nitrogen from tannery wastewater: performance and stability of a full-scale plant. *Journal of Chemical Technology and Biotechnology*, 79:879-888.

Third, K., Burnett, N., Cord-Ruwisch, R. (2003). Simultaneous nitrification and denitrification using stored substrate (PHB) as the electron donor in an SBR. *Biotechnology and bioengineering*, 83(6):706-720.

Valera, R. (2010). *Usos de amargos de salmueras como coagulantes para el tratamiento de aguas residuales.* Trabajo de Grado. Universidad del Zulia. [*Bitter brine used as coagulant for treatment of wastewaters.* Master Program in Environmental Science. Zulia University. Venezuela].

van Lier, J. B., Mahmoud, N., Zeeman, G. (2008). Anaerobic Wastewater Treatment. In: M. Henze, M. C. M. van Loosdrecht, G. A. Ekama, D. Brdjanovic (eds.), *Biological Wastewater Treatment, Principles, Modelling and Design.* IWA Publishing, London, Reino Unido. pp. 415-456. Chapter 16.

Vidal, G., Nieto, J., Cooman, K., Gajardo, M., Bornhardt, C. (2004). Unhairing effluents treated by an activated sludge system. *Journal of Hazardous Materials B*, 112:143–149.

Von Sperling, M., Lemos C. A. (2006). *Biological wastewater treatment in warm climate regions.* Londres, Reino Unido. 810 p.

Yildiz, G., Insel, G., Cokgor, E., Orhon, D. (2008). Biodegradation kinetics of the soluble slowly biodegradable substrate in polyamide carpet finishing wastewater. *Journal of Chemical Technology and Biotechnology*, 83:34-40.

In: Sequencing Batch Reactors: An Overview   ISBN: 978-1-53615-462-7
Editor: Lois K. Mello   © 2019 Nova Science Publishers, Inc.

Chapter 2

ASSESSMENT OF THE EFFECTIVENESS OF OULED BERJAL LANDFILL LEACHATE TREATMENT BY SEQUENTIAL BATCH REACTOR TECHNIQUE: ELIMINATION OF NITROGEN POLLUTION, PHENOLIC COMPOUNDS AND PHYTOTOXICITY

H. Bakraouy[1,*], S. Souabi[1], K. Digua[1], M. A. Bahlaoui[1],
A. Jada[2], H. Oubrayme[1], O. Dkhissi[1] and M. Chatoui[1]
[1]Process and Environmental Engineering Laboratory,
Faculty of Science and Technics, Hassan II University,
Mohammedia, Morocco
[2]Institute of Materials Science of Mulhouse, France

ABSTRACT

Solid waste management is one of the major issues faced by countries around the world. The waste management approach used in

* Corresponding Author's E-mail: h.bakraouy@gmail.com.

Morocco consists mainly to improve the conditions for waste collection, sorting, and treatment while limiting the nuisance caused by methane emissions and leachate production.

In the present work, we have treated leachate originated from Ouled Berjal landfill (Kenitra -Morocco) by using the Sequential Batch Reactor (SBR) technique. Thus, the leachate was put into two reactors, which differed in the times allocated to each phase. We have assessed the treatment effectiveness by monitoring the removal of NH_4^+ ions, nitrates, nitrites, phenol, absorbance at 254 nm and phytotoxicity. The results showed that the first SBR is suited for leachate treatment with removal efficiencies reaching 98.5% for NH_4^+ ions, 53% for nitrates, 61,5% for nitrites, 82,5% for phenol and 66% for the absorbance. In addition, the phytotoxicity tests showed an increase in the germination index from 19% to 47%, indicating that significant part of the pollutants was removed from the leachate.

Keywords: SBR, landfill leachate, phytotoxicity, phenol, nitrogen pollution

1. INTRODUCTION

Solid waste management remains a big challenge for countries around the world. In Morocco, the improvement in the standard of living, as well as the change in production and consumption patterns, led to an increase in the volume of solid waste, estimated at 6,9 million tones/year in 2015 (Report of the Ministry of the Environment, 2015). The quantity produced is 0,76 kg/day per habitant in urban areas and 0,3 kg/day per habitant in rural areas. Such increase affects both the quality of the collection, as well as, the quality of treatment in the landfill. Thus, the Landfills become sources of nuisance instead of being the solution to eliminate waste (Yong et al. 2018). They are sources of methane and leachate. Leachate is the result of the solubilization of compounds during the percolation of rainwater through the pile of waste, as well as moisture contained in the waste itself (Renou et al. 2008).

The challenge is now significant because it will be necessary to ensure effective drainage, recovery, and treatment of leachates for reuse or rejection in the nature. This approach will provide a significant water resource for use, particularly in areas experiencing water shortages and drought.

Due to their toxicity, we have to treat leachates before their release into the natural environment (Bashir et al. 2017).

Several studies on the characterization of landfill leachate revealed that it is a very complex medium containing more than 200 families of organic compounds.

According to Peng (2013), the pollutants contained in leachate can be classified in four categories:

- Dissolved or suspended organic matter such as chemical oxygen demand, volatile fatty acids, humic and fulvic compounds
- Mineral macros such as Sodium, Calcium, Magnesium, Iron, Ammonium ions
- Heavy metals such as Zinc, Cadmium, Nickel, Chrome
- Xenobiotic organic compounds such as aromatic hydrocarbons, phenols, chlorinated aliphatic hydrocarbons, pesticides and plasticizers

There are several leachate treatment techniques, namely: Leachate transfer, biological, physicochemical and membrane processes (Souabi et al. 2011).

The choice of a treatment technique in spite of another is strongly linked to the composition of the leachate, its age, but also to the purification efficiency required by the standards applied in a country.

Among the biological processes, we distinguish Sequential Batch Reactors (SBR). SBR consists in a discontinuous mixed biomass suspension culture process in which the biological reaction phases and settling processes take place in the same basin (Ismail et al. 2014).

The SBR process is used mainly to achieve the nitrification-denitrification of the effluent, which leads to the elimination of ammonium ions and their conversion into nitrites and nitrates, and finally to nitrogen gas (Liu et al. 2015).

The SBR treatment has several advantages compared to conventional activated sludge treatment. Indeed, in the SBR process, there are different growth kinetics due to the presence of different microorganisms species. Such different growth kinetics overcome the problem of appearance of filamentous bacteria (McQuarrie and Boltz, 2011). In addition, the SBR improves the reaction kinetics by ensuring substrate concentrations and maximum reaction rates at the end of the filling phase.

The SBR operating cycle consists of five steps, namely: Feeding, reaction, settling, draw, and idle. The succession of the reaction steps (which is characterized by alternating aeration and anoxic phases) and decantation lead to reduction in nitrogen, carbon and phosphorus pollution (Sathian et al. 2014).

Ranjan et al. (2016) carried out the co-treatment of landfill leachate and municipal wastewater in a SBR. Elimination efficiencies were 93% for NH_3, 83% for nitrates, 70% for COD, 80% for phosphorus and 83% for turbidity.

Sun et al. (2010) studied leachate treatment from Peking landfill in China using a UASB system combined with an SBR. Total nitrogen removal reached 95.4% at a temperature of 14.9°C.

The present work is devoted to the biological treatment of the leachate coming from Ouled Berjal landfill, through two SBRs processes that differ in the times allocated to each step. The effectiveness of the treatment was evaluated by monitoring the removal of ammonium ions, nitrates, nitrites, phenol, Absorbance at 254nm (Abs 254nm) and phytotoxicity.

2. Experimental Details

2.1. Sanitary Landfill Leachate Sample

We collected the raw leachate sample from the sanitary landfill of Ouled Berjal, located in Kenitra, Morocco. The rapid development of Kenitra city is followed by increased solid waste production, reaching about 120,000 tons in 2011 and 130,296 tons in 2015.

In Ouled Berjal, landfills are collected household and similar waste as well as green waste coming from Kenitra city. This waste is composed of 70% of organic matter.

This site covers a surface of about 20 ha, made of:

- Landfill area consisting of a 4 ha bin which is subdivided into 4 cells;
- Leachate and stormwater drainage pipe;
- Three large basins to collect leachate and a reserve basin:

Figure 1 illustrates Ouled Berjal landfill.

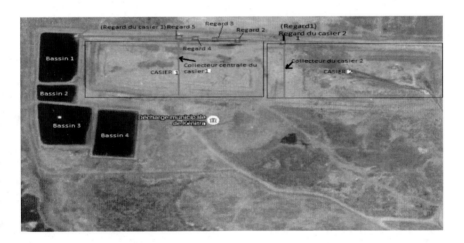

Figure 1. Ouled Berjal landfill.

Currently, Ouled Berjal landfill faces several difficulties. Indeed, there is a great risk of saturation of the landfill. In addition, leachates are stored in basins and are not treated. This can lead to overflowing leachates at any time, which can alter the groundwater and surrounding agricultural land.

Landfill leachate samples were collected in 30 liter plastic bottles, transported to the laboratory and analyzed before starting the SBR treatment.

Leachate was analyzed for phenol, turbidity, suspended matter, COD, BOD_5, color and Abs 254nm in accordance with the standard methods of the APHA (APHA, 2005), and for toxicity in accordance with the Wang method (Wang, 1989).

We have carried out the germination tests on 15 turnip seeds on petri dishes with filter paper soaked in raw or treated leachate. We have performed three replicates for each effluent (Wang, 1989). We conducted the tests in the dark and at room temperature (25°C) for a period of 120 hours. Further, we carried out the phytotoxicity test by monitoring the emergence of seeds, and the growth of radicles.

The germination index (GI) is determined with respect to emergence and root growth in leachate and distilled water according to the following equation:

$$GI(\%) = \frac{NG_{Lix} * LR_{Lix}}{NG_{ED} * LR_{ED}} * 100$$

with:

NG_{Lix}, NG_{ED}: Number of germinated seeds in leachate and distilled water.

LR_{Lix}, LR_{ED} : Length of radicles in leachate and distilled water.

2.2. Experimental Procedure

In this study, leachate from basin B_1 was treated biologically in an SBR for a period of 30 days. To do this, we have implemented two SBR reactors that differ in the duration allocated to the phases that constitute the operating cycle. The succession of phases and their durations are as follows (Figure 2).

The other operating conditions, such as the hydraulic residence time in the SBR, would be 30 h, while the cellular retention time (sludge age) would be maintained at approximately 15 days. The importance of using this last time is because high sludge ages allow the enrichment of the biomass of nitrifying autotrophic microorganisms, which are characterized by having a slower growth rate than heterotrophs, thus favoring the realization of the biological processes for the removal of nitrogen (Ekama and Wentzel, 2008a; Carrasquero, 2011).

- for SBR1 :

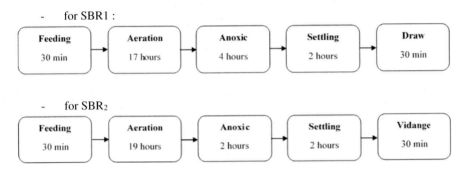

- for SBR2

Figure 2. Cycles adopted for the two SBRs during the experiment.

2.2.1. Sludge Age

We chose an infinite sludge age, i.e., the inoculum and the sludge formed during the biodegradation process were only removed at the end of the experimentation period (30 days). The inoculum used for this study consists of sludges from the STEP of Lessieur crystal company (a vegetable oil refinery industry). During the experimental period, the MLSS (Mixed Solids Suspended Solids) of the two SBRs reached 13,000 and 9,000 mg/L respectively for SBR_1 and SBR_2.

2.2.2. Dissolved Oxygen Content and Carbon Source

Aeration was provided by aquarium aerators linked to timers to adjust the duration and sequence of different phases.

The dissolved oxygen content was maintained at values around 2 mg/L. No carbon sources were added to the leachate during the experimental period.

2.2.3. Start-Up and Stabilization of SBR_1 and SBR_2

Both reactors were fed with 3 liters of water to be treated. They are kept under magnetic stirring during all phases except for settling and draw.

These two reactors were subjected to an acclimation phase that lasted 7 days; they were fed with a mixture of 50% leachate and 50% tap water. This phase aims to adapt the biomass of the inoculum to the new substrate, which is in our case landfill leachate.

From the eighth day, both reactors were fed with raw leachate.

2.2.4. Evaluation of Treatment Efficiency

A peristaltic pump "Perimatic GP" of Jencons brand carried out the draw (Figure 3). Treated leachate was analyzed daily in terms of the following parameters: COD, BOD_5, NH_4^+, NO_3^-, NO_2^-, turbidity, phenol, surfactant, total phosphorus, color and Abs 254nm.

In addition, we monitored the evolution of the phytotoxicity of treated leachates compared to raw leachate using the turnip grain germination method.

Figure 3. Draw the SBR through the peristaltic pump.

3. RESULTS AND DISCUSSIONS

3.1. Characteristics of the Landfill Leachate

The physicochemical analysis of the leachate is an essential step before starting the treatment. Indeed, the choice of the treatment technique to be used is strongly related to the composition. On the other hand, this analysis also makes it possible to determine the impact of leachate on groundwater (in case of infiltration), or on the environment adjacent to the landfill as well as on human health.

The composition of leachate depends on several parameters including: The age of the landfill, type of waste and the technique used for landfilling (Chofqi et al. 2007).

The physicochemical characteristics of the leachates used for this study are summarized in Table 1.

We note that the leachate from B_1 basin is loaded with organic matter, as evidenced by the high COD value of 11,5 g/L. This remains below the values reported by Ezzoubi et al. (2010) and Jirou et al. (2014), which are 54 g/L and 72 g/L, respectively.

Table 1. Mean and extent of the different physicochemical parameters of the studied leachate

Parameter	Mean	Max	Min
Turbidity (NTU)	110,8	149	75,2
Phenol (mg/L)	158,6	409,6	69,8
NO_3^- (mg/L)	356,0	416,0	248,0
NH_4^+ (mg/L)	1095,3	1231,5	523,5
NO_2^- (mg/L)	716,7	1040,0	460,0
Total Phosphorus (mg/L)	26,2	33,2	16,9
Surfactant (mg/L)	55,0	67,7	16,5
COD (mg O_2/L)	7840,0	11520	3840
Color ($F_D = 200*$)	0,039	0,057	0,025
Abs 254nm ($F_D = 200*$)	0,192	0,380	0,073

*F_D = Dilution factor.

This leachate can be classified as young, since the value of the COD exceeds 10 g/L (Renou et al. 2008).

It contains 1,1 g/L of NH_4^+, which is close to the value reported by Zhu et al. (2013), for leachate from the city of Beijing.

The mean nitrate concentration reached 356 mg/L. In a research conducted by Tahiri et al. (2014) on the Meknes City landfill, the leachate characterization revealed a nitrate content of 751,1 mg L, which is higher than the present study. In contrast, the concentration of nitrates is lower in the leachates of El Jadida city landfill: 2,4 mg/L (Chofqi el al. 2007), and Fez city landfill: 11,3 mg/L (Ezzoubi et al. 2010), as well as the city of Oran: 0,92 mg/L (Bennama et al. 2010).

The mean concentration of total phosphorus is 26,2 mg/L, which exceeds the values reported by Silva et al. (2013), with a value of 19,1 mg/L, and by Zhu et al. (2013), with a value of 5,2 mg/L, but remains lower than that of the Agadir city landfill, which reached 188,4 mg/L (Jirou et al. 2014).

3.2. Elimination of Ammoniacal Nitrogen

In addition to refractory organic matter, landfill leachate contains high levels of ammonium. According to Jokela and Rintala (2003), 50% of the nitrogen is solubilized following the anaerobic digestion of municipal solid waste. The fermentation of the amino acid leads to the formation of organic acids and ammonia. Thus, due to the hydrolysis and fermentation of these nitrogen fractions, leachate from old landfills is heavily loaded with ammonia nitrogen (Kulikowska and Bernat, 2013).

One of the major goals of SBR treatment is the elimination of nitrogen pollution. In fact, in an SBR, the nitrification-denitrification process takes place, which consists initially in the transformation of ammoniacal nitrogen into nitrates and nitrites (nitrification) and then in gaseous nitrogen (denitrification).

We consider NH_4^+ removal as an indicator of the completion of the acclimation phase.

Indeed, Figure 4 shows that during the first days, elimination of NH_4^+ of treated leachates is low.

This is probably due to the non-adaptation of the bacteria. Indeed, from the 5th day, we note that removal of NH_4^+ stabilized at 35% for SBR_1 and 30% for SBR_2, which means that the acclimation phase took end. Therefore, we can feed the two reactors with raw leachate without dilution and this from the 8th day. Moreover, NH_4^+ concentration at the outlet of the two SBRs continues to decrease progressively to reach minimum concentrations of 7,9 and 7,5 mg/L for SBR_1 and SBR_2 respectively, achieving elimination efficiencies exceeding 98%.

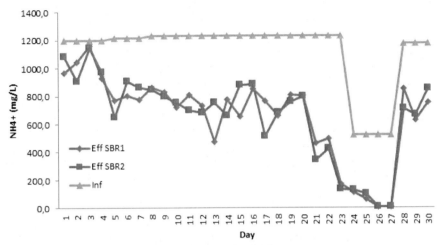

Figure 4. Variation of NH_4^+ concentration during SBR treatment.

These efficiencies are higher than that achieved by Neczaj et al. (2005). Indeed, the study of these authors focused on the treatment of the stabilized leachate by ultrasound coupled to SBR. The cycle time of the SBR is 3 hours, divided as follows: Feeding (0,15 h), reaction (3h), decantation (0,3 h), draw (0,08 h). As a result, 90% of NH_4^+ was eliminated.

3.3. Elimination of Nitrites and Nitrates

The formation of nitrites and nitrates, commonly known as nitrification, is obtained aerobically via the following reactions:

$$NH_4^+ + \frac{3}{2}O_2 \rightarrow NO_2^- + H_2O + 2H^+ \text{ (Nitritation)}$$

$$NO_2^- + \frac{1}{2}O_2 \rightarrow NO_3^- \text{ (Nitratation)}$$

Nitritation requires a reduction of the activity of oxidative nitrite bacteria (ONB), without affecting oxidative ammonia microorganisms (OAB). This can be achieved by limiting the dissolved oxygen supply as well as regulating the pH (Kulikowska and Bernat, 2013).

In addition, the elimination of these two elements (nitrates and nitrites) is commonly called denitrification, which is carried out under anaerobic conditions. It is governed by the following reaction:

$$NO_3^- \rightarrow NO_2^- \rightarrow NO \rightarrow N_2O \rightarrow N_2$$

Figure 5 shows that both SBRs can remove up to 30% of the nitrites. However, we notice that the concentration of nitrites at the exit of SBR is higher than that at the inlet, which indicates that there is a significant elimination of NH_4^+. As a result, there is a formation of nitrites, which are less toxic compared to NH_4^+.

As a result, we were unable to achieve complete denitrification. This is mainly due to the lack of organic carbon source, since our experiments were conducted without resorting to an external carbon source. Indeed, Tora et al. (2011) indicated that the COD/N ratio required for the complete elimination of nitrites depends on the type of organic carbon source (ethanol, glycerin, etc.). This ratio must be equal to 3 if ethanol is added as a carbon source. In addition, the kinetics of conversion of NH_4^+ to NO_3^- and NO_2^- is often slow.

The concentration of nitrates (Figure 6) marks a trend that fluctuates a lot. In general, it is higher at the entrance than at the exit of the two SBRs, except for the period from the 12th to the 25th day. Which shows that denitrification is implemented.

Removal efficiencies of nitrates reached 53,1% for SBR_1 and 51% for SBR_2.

Kulikowska and Bernat (2013) investigated the effect of limited oxygen alimentation during the aeration phase of the SBR cycle on the denitrification efficiency of landfill leachate. The mixture of glycerin with Sodium acetate or Sodium acetate was used as an external carbon source. These authors showed that 99% of the ammonium ions and 61% of the nitrites were eliminated.

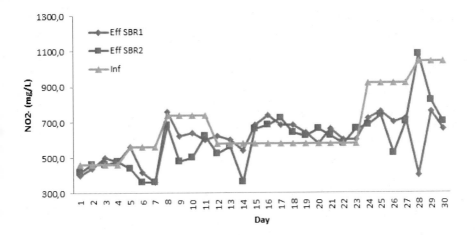

Figure 5. Variation of NO_2^- concentration during SBR treatment.

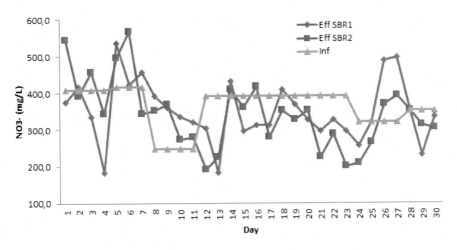

Figure 6. Variation of NO_3^- concentration during SBR treatment.

3.4. Elimination of Phenol and Abs 254nm

Through Figures 7 and 8, we notice that the variations of phenol and Abs 254nm elimination efficiencies follow the same trend.

We note that removal of phenol and Abs at 254 nm are low during the first 11 days of testing. This may be due to the formation of complex molecules from other simpler ones, or to the biodegradation of non-humic compounds followed by the formation of poly-compact humic structures (Bu et al. 2010). As a result, removals increase (the two curves follow the same trend) to reach:

- 82,5 and 80,5% for removal of phenol for respectively for SBR_1 and SBR_2;
- 65,8 and 68,2% for removal of Abs 254nm respectively for SBR_1 and SBR_2.

Yussof et al. (2016) treated a synthetic industrial wastewater by an SBR operating for 24h cycle, distributed as follows: 1h of feeding, 20h of aeration, 2h of settling and 1h of drawing. As a result, the phenol content was reduced by 80%.

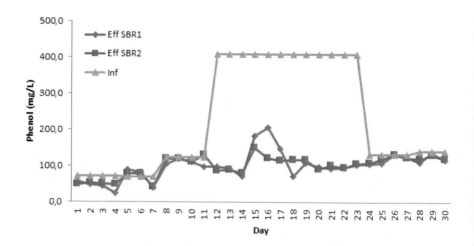

Figure 7. Variation of phenol concentration during SBR treatment.

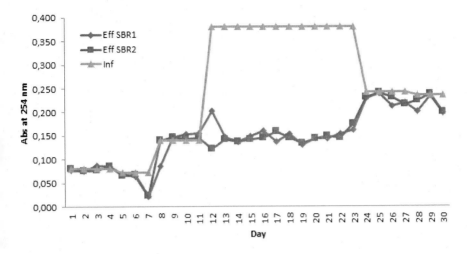

Figure 8. Variation of Abs 254nm concentration during SBR treatment.

3.5. Reduction of Phytotoxicity

Landfill leachate is a real danger to people and the environment. This is due to the high concentrations of recalcitrant pollutants whose toxicity is very worrying. Therefore, the reduction of leachate toxicity is an important factor in the treatment of this effluent (Silva et al. 2004).

Phytotoxicity tests make it possible to highlight the impact of leachate on the soil, in particular on the germination of grains.

Phytotoxicity is often assessed by the study of germination or by growth tests (Wang et al. 2001). Several authors used phytotoxicity tests to evaluate the toxicity of raw and treated leachates and thus the effectiveness of the treatment. In fact, treated leachate must not contain substances that prevent seed germination and plant growth. Several seeds can be chosen for this test in view of their sensitivities (Turnip, Watercress, Alfalfa, Tomato, Lettuce ... etc.).

In this study, we opted for the method of germination of turnip seeds, as it is a very sensitive species. Thus, the evaluation of phytotoxicity is carried out by calculating the germination index and the length of the radicles.

H. Bakraouy, S. Souabi, K. Digua et al.

We compared raw leachates with treated leachates during the 12th and 21st days to evaluate the efficacy of SBR_1 and SBR_2 treatment.

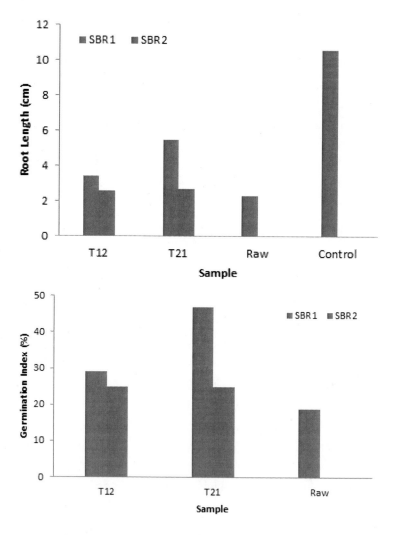

Figure 9. Evaluation of phytotoxicity parameters for leachates treated with SBR.

Referring to Figure 9, we note that germination indices and root lengths for both SBRs are higher than raw leachate. This shows that the toxicity of treated leachates is lower than that of raw leachate due to the elimination of toxic elements.

The germination index is 19% for raw leachate. For SBR_1, this index increases from 29% (for the 12th day) to 47% (for the 21st day). Comparing the leachates treated during the 12th and the 21st day, we note an increase in the length of the radicles of the order of 2.1 cm.

For SBR_2, the germination index reached 25%, it is higher than the raw. However, no improvement was observed following the comparison of the two leachates treated during the 12th and 21st days.

It follows that in terms of phytotoxicity, SBR_1 is better suited for the elimination of toxic compounds contained in leachate coming from B_1 basin.

Olivero-Verbel et al. (2008) evaluated the phytotoxicity of leachate using the aquatic species Artemia franciscana. The purpose of the study is to determine the relationship between composition and toxicity of leachate. The authors showed that the toxicity decreases with the increase of ionic substances. In contrast, organic matter (COD) and heavy metals such as cadmium act as a promoter of toxicity. The authors conclude that toxicity may depend on the formation of organic complexes rather than free ionic compounds.

Li et al. (2010) studied the evolution of the toxicity of stabilized leachate before and after coagulation-flocculation treatment. They chose the wheat seed growth inhibition test because wheat is the main grain consumed around the world. At the end of this study, samples treated with aluminum-based coagulants showed the highest toxicity. This has been explained by the existence of residual aluminum (in the form of Al^{3+} ions) in the treated leachate, which is toxic to plants. The authors add that the toxicity is strongly related to the organic matter contained in the leachate, since the toxicity increases at high concentrations of COD.

CONCLUSION

This study showed the effectiveness of the Sequential Batch Reactor (SBR) technique in the treatment of leachate from Ouled Berjal landfill. Thus, we have compared two SBRs, which differ by the time allocated to the different phases of the operating cycle. We have obtained the best results with SBR having the following cycle phases: Feeding (30 min), Aeration (17 hours), Anoxic (4 hours), Settling (2 hours) and draw (30 min).

The observed removal efficiencies were: 98,5% of NH_4^+, 53% of nitrates, 61,5% of nitrites, 82,5% of phenol and 66% of the absorbance at 254nm. We also examined the efficiency of SBR treatment on the phytotoxicity reduction. Hence, we conducted phytotoxicity tests that consist of monitoring germination indices and root lengths of turnip seeds. According to the data, we concluded that SBR_1 is more efficient in phytotoxicity reduction, since the germination index increases from 19% to 47%.

ACKNOWLEDGMENTS

The authors are thankful to Moroccan Ministry of Environment who funded this study.

REFERENCES

Bashir, M. J. K., Xian, Tay Ming, Shehzad, A., Sethupathi, S., Ng, C. A., Abu Amr, S., 2017. Sequential treatment for landfill leachate by applying coagulation-adsorption process. *Geosyst. Eng.* 20, 9-20.

Bennama T., Younsi A., Derriche Z. and Debab A., Caractérisation et traitement physico-chimique des lixiviats de la décharge publique d'El-Kerma (Algérie) par adsorption en discontinu sur de la sciure de bois naturelle et activée chimiquement, *Water Qual. Res. J. Can.* 45 (1), pp. 81–90. (2010). [Characterization and physicochemical treatment of leachates of the El-Kerma dump (Algeria) by batch adsorption on natural and chemically activated sawdust]

Chofqi, A., Younsi, A., Lhadi, E. K., Mania, J., Mudry, J., Veron, A. Lixiviat de la décharge publique d'El Jadida (Maroc): Caractérisation et étude d'impact sur la nappe phréatique, *Déchets sciences et techniques*, N°46, pp. 4-10, (2007). [Leachate of the El Jadida landfill (Morocco): Characterization and impact study on groundwater, *Waste science and technology*]

Ez zoubi Y., Merzouki M., Bennani L., El Ouali Lalami A. and Benlemlih M., Procédé pour la réduction de la charge polluante du lixiviat de la décharge contrôlée de la ville de Fès, *Déchets, sciences et techniques - Revue francophone d'écologie industrielle* 58, pp. 22-29 (2010). [Process for the reduction of the pollution load of the leachate of the controlled landfill of the city of Fes, *Waste, science and technology - French-speaking review of industrial ecology*]

Ismail, N. F., Jami, M. S., Amosa, M. K., Muyibi, S., 2014. Modelling and simulation of energy-saving potential of sequential batch reactor (SBR) in the abatement of ammoniacal-nitrogen and organics. *J. Pure Appl. Microbiol.* 8, 809-814.

Jirou Y., Harrouni M. C., Belattar M., Fatmi M. and Daoud S., Traitement des lixiviats de la décharge contrôlée du Grand Agadir par aération intensive, *Rev. Mar. Sci. Agron. Vét.* 2 (2), pp. 59-69 (2014). [Treatment of leachate from the Grand Agadir controlled landfill by intensive aeration]

Jokela J. P. Y. and Rintala, J. A., Anaerobic solubilisation of nitrogen from municipal solid waste (MSW), *Rev. Environ. Sci. Biotechnol.* 2, pp. 67–77 (2003).

Kulikowska D. and Bernat K., Nitritation–denitritation in landfill leachate with glycerine as a carbon Source, *Bioresource Technology* 142, pp. 297–303 (2013).

Li, W., Hua, T., Zhou, Q. X., Zhang, S. G., Li, F. X., 2010. Treatment of stabilized landfill leachate by the combined process of coagulation/flocculation and powder activated carbon adsorption. *Desalination* 264, 56–62.

Liu, Z., Wu, W. H., Shi, P., Guo, J., Cheng, J., 2015. Characterization of dissolved organic matter in landfill leachate during the combined treatment process of air stripping, Fenton, SBR and coagulation. *Waste Manage.* 4, 111–118.

McQuarrie J. P., Boltz J. P., Moving bed biofilm reactor technology: Process applications, design, and performance, *Water Environ. Res.* 83 (2011) 560–575.

Neczaj E., Okoniewska E. and Kacprzak M., Treatment of landfill leachate by sequencing batch reactor, *Desalination* 185, pp. 357–362 (2005).

Olivero-Verbel J., Padilla-Bottet C., De la Rosa O., Relationships between physicochemical parameters and the toxicity of leachates from a municipal solid waste landfill, *Ecotoxicology and Environmental Safety* 70, pp. 294–299 (2008).

Peng, Y. (2013). Perspectives on technology for landfill leachate treatment. *Arabian Journal of Chemistry,* 10, S2567–S2574.

Ranjan K., Chakraborty S., Verma M., Iqbal J. and Naresh Kumar R., Co-treatment of old landfill leachate and municipal wastewater in sequencing batch reactor (SBR): effect of landfill leachate concentration, *Water Quality Research Journal of Canada* 51(4), pp. 377-387 (2016).

Renou, S., Givaudan, J. G., Poulain, S., Dirassouyan, F., Moulin, P., 2008. Landfill leachate treatment: review and opportunity. *J. Hazard. Mater.* 150, 468–493.

Sathian, S., Rajasimman, M., Radha, G., Shanmugapriya, V., & Karthikeyan, C. (2014). Performance of SBR for the treatment of textile dye wastewater: Optimization and kinetic studies. *Alexandria Engineering Journal,* 53(2), 417–426.

Silva A. C., Dezotti M., G. L. and Jr. Sant'Anna, Treatment and detoxification of a sanitary landfill leachate, *Chemosphere* 55, pp. 207–214 (2004).

Silva A. C., Dezotti M., G. L. and Jr. Sant'Anna, Treatment and detoxification of a sanitary landfill leachate, *Chemosphere* 55, pp. 207–214 (2004).

Souabi, S., Touzare, K., Digua, K., Chtioui, H., Khalil, F., Tahiri, M., 2011. *Sorting and Valuation Solid Waste in the Garbage Dump of the*

City of Mohammedia, Thesis in Chemistry, vol. 6. Faculty of Science and Technical Chemistry Department Fes, No. 25.

Sri Shalini S. and Joseph K., Nitrogen management in landfill leachate, Application of SHARON, ANAMMOX and combined SHARON–ANAMMOX process. *Waste Manage*. 32, pp. 2385–2400 (2012).

Sri Shalini S. and Joseph K., Nitrogen management in landfillleachate, Application of SHARON, ANAMMOX and combined SHARON–ANAMMOX process. *Waste Manage*. 32, pp. 2385–2400 (2012).

Sun H., Yang Q., Peng Y., Shi X., Wang S. and Zhang S., Advanced landfill leachate treatment using a two-stage UASB-SBR system at low temperature, *Journal of Environmental Sciences* 22(4), pp. 481–485 (2010).

Tahiri A. A., Laziri F., Yachaoui Y., El Jaafari S. and Tahiri A. H., Etude des polluants contenus dans les lixiviats issus de la décharge publique de la ville de Meknès (MAROC), *European Scientific Journal* 10(35), pp. 170-186 (2014).

Third report on the state of the environment of Morocco, Ministry Delegate to the Minister of Energy of Mines, Water and Environment, in charge of the environment, (2015).

Tora J. A., Baeza J. A., Carrera J. and Oleszkiewicz J. A., Denitritation of a high strength nitrite wastewater in a sequencing batch reactor using different organic carbon sources, *Chem. Eng. J.* 172, pp. 994–998 (2011).

Wang W., Toxicity assessment of the aquatic environment using phytoassay methods. *By Water Quality Section Illinois State Water Survey* (1989).

Wang X., Sun C., Gao S., Wang L. and Shuokui H., Validation of germination rate and root elongation as indicator to assess phytotoxicity with *Cucumis sativus*, *Chemosphere* 44(8), pp. 1711-21 (2001).

Yong Z. J., M. J. K. Bashir, C. A. Ng, S. Sethupathi, J. W. Lim. A sequential treatment of intermediate tropical landfill leachate using a sequencing batch reactor (SBR) and coagulation. *Journal of Environmental Management* 205 (2018) 244-252.

Yusoff N., Ong S., Ho L. N, Wong Y. S., Mohd Saad F. N., Khalik W., Lee S. L. (2016). Evaluation of biodegradation process: Comparative study between suspended and hybrid microorganism growth system in sequencing batch reactor (SBR) for removal of phenol. *Biochemical Engineering Journal* 115 (2016) 14–22.

Zhu R., Wang S., Li J., Wang K., Miao L., Ma B. and Peng Y., Biological nitrogen removal from landfill leachate using anaerobic–aerobic process: Denitritation via organics in raw leachate and intracellular storage polymers of microorganisms, *Bioresource Technology* 128, pp. 401–408. (2013)

In: Sequencing Batch Reactors: An Overview   ISBN: 978-1-53615-462-7
Editor: Lois K. Mello   © 2019 Nova Science Publishers, Inc.

Chapter 3

ANSBBR APPLIED TO BIOMETHANE PRODUCTION BY THE TREATMENT OF VINASSE, WHEY AND GLYCERIN: EFFECTS OF MESOPHILIC AND THERMOPHILIC CONDITIONS

J. N. Albuquerque, S. P. Sousa, G. Lovato, R. Albanez, S. M. Ratusznei and J. A. D. Rodrigues[*]

Mauá School of Engineering, Mauá Institute of Technology
(EEM/IMT), São Caetano do Sul, SP, Brazil

ABSTRACT

Anaerobic digestion allows organic matter conversion into final products such as methane by microorganism activity. Over the last decades, engineering has adapted this process into anaerobic bioreactors

[*] Corresponding Author's E-mail: rodrigues@maua.br.

with different configurations for wastewater treatment aimed at biogas production. Among reactor types, the anaerobic sequencing batch reactor (ASBR) is one of the several high rate configurations and it appears as an alternative to continuous systems. It presents advantages such as better effluent control and simple operation comprised in four steps: feed, reaction, settling of granular biomass and decant. Another configuration consists in the anaerobic sequencing batch biofilm reactor (AnSBBR), in which the biomass is immobilized in inert support. It enables the elimination of the settling step, thus reducing the overall cycle time. Although they have different biomass arrangements, the operational factors tend to be the same: agitation, food/microorganism ratio, reactor configuration and feed strategy. AnSBBRs have been applied to mesophilic and thermophilic treatment of wastewaters from agroindustry such as vinasse (bioethanol production), whey (dairy industry), and glycerin (biodiesel production) with various operational strategies: feeding mode, temperature, organic load, influent concentration and cycle time. Therefore, this study presents an overview of achievements of studies that used AnSBBRs digesting agroindustry wastes for methane production, focused on operational strategy and perspectives for scale-up estimative.

Keywords: anaerobic digestion, ASBR, AnSBBR, methane, mesophilic, thermophilic

1. INTRODUCTION

The agroindustry wastewater is problematic for the environment due to its composition, pollutants concentration and volumes generated. For example, in Brazil, the biodiesel industry has glycerol as the major by-product of the biodiesel industry, which in general corresponds to 10-18% by mass of biodiesel produced. Pure glycerol has applications in many industries, but the refinement of crude glycerol to a high purity is too expensive, especially for small and medium biodiesel producers. In ethanol production process generates large volumes of effluents, mainly vinasse that is generated in the proportion of 12 to 15 L per liter of ethanol produced. Another example is in the dairy industry, whey represents about 90% of the volume of milk used in cheese production. Therefore, these residues have become a challenge for the agribusiness, since if disposed without any type of treatment can lead to serious damages to the environment (Lovato et al., 2016 and Albanez et. al., 2016a). Therefore, these residues have become a challenge for the agribusiness, since if disposed without any type of treatment can lead to serious damages to the environment.

Biological treatment is a reliable method to treat these types of wastewaters. There are two main types of biological treatment, depending on the oxygen requirements: aerobic and anaerobic methods. The aerobic treatment, despite the high efficiency of organic matter removal, demands a high energy consumption to perform the air injection by the aerators, besides causing a high generation of sludge that could become another problem to be treated. On the other hand, anaerobic treatment has good removal efficiency, low energy consumption, low sludge generation, and biogas production, which can result in energy recovery (Cakir and Stenstromb, 2005).

The anaerobic digestion of substrates with high organic load is usually performed in high rate anaerobic reactors, among which anaerobic reactors operated in sequential batch with immobilized biomass (AnSBBR) have been outstanding due to the high performance and stability standards achieved, besides flexibility of operation and allow greater control over the effluent quality (Lovato et al., 2016).

AnSBBR has been the focus of several studies aiming the biotechnological feasibility related to aspects of environmental suitability and production of biogas from several types of wastewater, in order to allow its scale-up application. (Manssouri et al., 2013). The performance of the AnSBBR is directly related to some operational aspects, such as inert material for biomass immobilization, feed strategy, type of substrate, organic matter concentration in the wastewater, type of agitation, cycle time and temperature (Albanez et al., 2016a).

In this scenario, this work presents studies that used AnSBBR in order to evaluate the influence of operational aspects on methane production from the anaerobic digestion of glycerin, whey or vinasse, in both conditions - mesophilic and thermophilic, considering the performance indicators of operational stability and biogas productivity. It was also analyzed the relationship between methane produced and organic matter consumed, relating in this way fundamental and technological aspects for an evaluation of the use of this technology in real scale.

2. ANAEROBIC SEQUENCING BATCH/FED-BATCH BIOFILM REACTOR (AnSBBR)

Discontinuous processes, such as anaerobic reactors operated in sequential batches (ASBR), offer a series of advantages in relation to continuous processes, such as better operational control, higher solids retention, high efficiency of organic matter removal, efficient effluent quality control and the possibility of application to a large variety of wastewaters (Sung and Dague, 1995; Zaiat et al., 2001). Their operational flexibility makes it viable their application to industries with intermittent effluent disposal, very restrictive emissions standards, recalcitrant compounds and systems that seek reuse or recovery of dissolved compounds (Zaiat et al., 2001).

The typical cycle of the ASBR comprises four steps: *(i)* feeding with different feeding times, which defines the feeding strategy as batch or fed-batch; *(ii)* the treatment itself or reaction time, in which the wastewater is converted to biogas by microorganisms; *(iii)* settling if the biomass is in granules (ASBR), but when the biomass is immobilized in inert support (AnSBBR), this step is not necessary; and *(iv)* discharge of the treated and clarified effluent, which allows the reactor to be fed again, therefore initiating a new cycle (Sung and Dague, 1995).

Biomass arrangement is fundamental for reactor operation and performance. Its arrangement can be in granular form or immobilized in inert support. Biomass immobilization in this type of reactor allows the elimination of the settling step and, therefore, the reduction of cycle time plus. It also prevents the discharge of biomass with the treated effluent. The use of an anaerobic sequencing batch reactor with immobilized biomass (AnSBBR) was initially demonstrated by Hirl and Irvine (1996), with biomass immobilized in gravel and homogenization performed by liquid phase recirculation. Another model was later described by Ratusznei et al., (2000) in which biomass was immobilized in cubes of polyurethane foam, arranged in a basket inside the reactor, and the stirring was performed mechanically by magnetic bars. A more in-depth analysis of these two configurations (ASBR and AnSBBR) can be found in the literature review by Lovato et al., (2015).

The key aspects of anaerobic sequencing batch reactors with immobilized biomass are medium agitation, feeding strategy, food/microorganisms ratio (S/X) and organic loading rate (Zaiat et al., 2001).

2.1. Agitation

Agitation is required during the reaction step, it improves mass-transfer velocities and ensures homogeneous conditions of temperature, pH and substrate concentration inside the reactor, essential factors for good system performance (Sung and Dague, 1995). Mixing can be performed by mechanical agitation, recirculation the liquid phase or by gas phase recirculation, the latter being inadequate in cases of low biogas production, such as in processes with low organic loading rate (Zaiat et al., 2001).

Damasceno et al., (2008) studied the effect of impeller type (turbine or propeller) and frequency (100, 200, 300 and 500 rpm) in an AnSBBR treating whey-based wastewater. There was no difference between impeller types when using higher frequencies, but the propeller type featured greater stability when with smallest frequencies. Novaes et al., (2010) reported an improvement in substrate consumption velocity and in solubilization of particulate organic matter when frequency of rotation was increased from 40 to 80 rpm; in addition, the authors studied different types of impellers, leading to an observation that the propeller type presented advantages over the flat and tilted blade impellers. Cubas et al., (2011) studied impeller types (propeller, flat and curved blade turbines) and rotation frequencies (from 100 to 1100 rpm) in an AnSBBR treating synthetic wastewater at low organic loading rate, the best results were achieved using the flat blade turbine. In addition, the authors concluded that the mass transfer in liquid phase was affected not only by the rotation frequency, but also by the speed imposed on each type of impeller.

Camargo et al., (2002) compared the performance of an AnSBBR reactor with and without liquid phase recirculation treating low strength wastewater, the recirculation velocities were varied from 0 to 0.188 cm.s^{-1}. The authors reported a 12% increase in organic matter removal efficiency (83-95%) and a 40% reduction in the amount of volatile acids in the effluent when liquid phase recirculation velocity was of 0.03 and 0.09 cm.s^{-1}. Recirculation velocity in an AnSBBR was also studied by Ramos et al., (2003). The authors reported difference in COD removal efficiency from 72 to 87% for the reactor without and with recirculation, respectively. As for the influence of velocity, an increase in the kinetic parameter of 1.19 to 2.00 h^{-1} was reported with the increase in recirculation velocity from 0.03 to 0.19 cm.s^{-1}, which remained stable at higher speeds.

2.2. Feeding Strategy

The feeding strategy is directly linked to the substrate/biomass ratio (S/X), therefore using a suitable filling time allows a better system stability, since this parameter can reduce the impact of high organic loading rates, toxicity of compounds and absorb peaks of volatile acids (Zaiat et al., 2001; Rodrigues et al., 2011). While the batch mode achieves higher concentrations of biogas, the fed batch mode allows the maintenance of low concentrations of the compounds in the reaction medium and thus decreases the effects of inhibition of substrate degradation (Bagley and Brodkorb, 1999).

In studies of the influence of feeding strategy on AnSBBRs, Damasceno et al., (2007) analyzed the influence of filling time of 2 and 4 h for cycle times of 8 h and verified that in conditions operated at low organic loading rates (2 and 4 gCOD.L^{-1}.d^{-1}) the feeding time of 2 h achieved better results. On the other hand, when the organic loading rate was increased to 8 and 12 $gCOD.L^{-1}.d^{-1}$ the best results were obtained with the filling time of 4 h.

Novaes et al., (2010) reported that AnSBBRs present a lower sensitivity of exposed biomass when operated in fed batch fed compared to ASBRs, which may be advantageous when there are shock loads or system failure that can lead to biomass loss. Lovato et al., (2012) evaluated the effect of feeding time (2, 4 and 6 h) on methane production by treating effluent from biodiesel production and concluded that, despite the lower efficiency of organic matter removal, the 2 h condition of feed time is more favorable for methane production. Volpini et al., (2018) in mesophilic condition with AVOL of 8.3 $gCOD.L^{-1}.d^{-1}$ and Albuquerque et al., (2019) in thermophilic condition with AVOL of 25 $gCOD.L^{-1}.d^{-1}$ reported improvements in reactor efficiency regarding organic matter removal, acids control and methane productivity when operated in fed batch.

On the other hand, both Almeida et al., (2017), who treated vinasse in mesophilic condition at AVOL of 9.7 $gCOD.L^{-1}.d^{-1}$, and Siqueira et al., (2018), who treated cheese whey in thermophilic condition at AVOL of 31.9 $gCOD.L^{-1}.d^{-1}$, reported a worse reactor performance when the feed strategy was modified from batch (10 min of filling time with cycle time of 480 min) to fed batch (filling time of 240 min with cycle time of 480 min). These results were attributed to the high organic loading rate and some poorly soluble solids that hindered mass transfer inside the reactor.

2.3. Organic Loading Rate

The applied organic loading rate has relevant influence in process performance, since the amount of substrate available for microorganisms is directly related to acid production and consequently to methane production (Bezerra et al., 2009; Lovato et al., 2017; Albuquerque et al., 2019).

Volpini et al., (2018) studied the increase in the organic loading rate (1.9-8.8 $gCOD.L^{-1}.d^{-1}$) treating hydrogen reactor effluent (treating vinasse) in an AnSBBR. They reported improved productivity of methane and reduction in organic matter removal efficiency. Lovato et al., (2015) and

Albuquerque et al., (2019) also verified improvement in methane productivity with the increase of organic loading rate, such that the best results were achieved when the highest organic loading rate was applied. On the other hand, Almeida et al., (2017) also reported improved methane productivity as the organic loading rate increased (1.1-10.1 $gCOD.L^{-1}.d^{-1}$), but the best results were achieved at 8.2 $gCOD.L^{-1}.d^{-1}$, which means that the process had already reached its optimum AVOL.

2.4. Final Considerations

In literature, the use of ASBRs and AnSBBRs for methane and hydrogen production from various wastewaters, such as sanitary sewage, automobile industry, ethanol production (vinasse), biodiesel production (glycerin), dairy industry (cheese whey), brewing, among other industrial effluents (Bezerra et al., 2009 e 2011; Ribas et al., 2009; Oliveira et al., 2009; Rodrigues et al., 2011; Lovato et al., 2012; Albanez et al., 2016a; Lima et al., 2016; Almeida et al., 2017) has become more common.

Table 1. Studies that used ASBR/AnSBBR for wastewater treatment

Study objective	Substrate	Reference
Effect of surface velocity on mass transfer	Low strength wastewater	Ramos et al., (2003)
Influence of feeding time	Low strength wastewater	Borges et al., (2004)
Influence of recirculation	Hypersaline	Mohan et al., (2007)
Effect of organic load, shock load and alkalinity	Cheese whey	Bezerra et al., (2007)
Influence of organic loading rate increase	Vinasse	Ribas et al., (2009)
Sulfate removal	Domestic sewage	Archilha et al., (2010)
Effect of feeding strategy and organic loading rate	Industrial wastewater	Rodrigues et al., (2011)
Effect of feeding strategy and organic loading rate on methane production	Glycerin	Lovato et al., (2012)
Methane production, environmental compliance and scale-up estimation	Molasses and vinasse	Albanez et al., (2016a)
Effect of organic loading rate, feed strategy and temperature on methane production	Vinasse	Almeida et al., (2017)
Effect of organic loading rate and feed strategy on methane production	Vinasse	Albuquerque et al., (2019)

Table 1 shows some studies that used these reactors for the treatment of different types of wastewaters.

3. Treatment of Glycerin Based-Wastewater in Mesophilic and Thermophilic Conditions

Biodiesel has emerged as an attractive alternative to decrease the use of fossil fuels. In Brazil, since 2008, the addition of biodiesel on common diesel composition is required, it reached 3% in the second half of that same year and a gradual increase to 20% is expected until 2020. This incentive through Federal Law n° 11.097/2005 boosted the biodiesel market. According to the National Agency of Petroleum, Natural Gas and Biofuels (ANP) since the mandatory addition of biodiesel in the diesel mixture there was a significant increase of approximately 3.7 in generated volume, reaching more than 4,291,000 cubic meters produced by 2017. In the first two months of 2018 only a 1.3 increase was recorded.

Biodiesel is produced from vegetable oils through its transesterification with methanol, a process that consists of a chemical reaction catalyzed by an acid or a base (sodium hydroxide or potassium hydroxide), in this stage the main molecules of the oils are separated in fatty acids and glycerin (Larsen, 2009; Baba, 2013). Glycerol is the major byproduct of the biodiesel industry. In general, for every 100 kg of biodiesel produced, approximately 10 kg of crude glycerol is generated. Crude glycerol that is generated by homogeneous base-catalyzed transesterification contains approximately 50-60% glycerol, 12-16% alkalis, especially in the form of alkali soaps and hydroxides, 15-18% methyl esters, 8-12% methanol, and 2-3% water. In addition to methanol and soaps, crude glycerol also contains a variety of elements such as Ca, Mg, P, and S as well as other components (Rivero et al., 2014).

Despite the wide applications of pure glycerol in the pharmaceutical, food, and cosmetic industries, the refinement of crude glycerol to a high purity is too expensive, especially for small and medium biodiesel producers. Alternate ways of using the crude glycerol phase have recently been studied. Possibilities such as combustion, co-burning, composting, animal feed, thermochemical conversion, and biological conversion have been applied to crude glycerol processing (Johnson and Taconi, 2007; Mendes et al., 2012).

Among these different options, the biological production of methane from crude glycerol by anaerobic digestion has several advantages. In addition to the production of methane, the advantages include low nutrient requirements, energy savings, it is a substrate readily biodegradable, has a pH suitable for anaerobic processes and generation of a stabilized digestate that improves soil quality (Rivero et al., 2014).

3.1. Mesophilic Treatment

There are several works that present glycerin anaerobic treatment, both in hydrogen production (Ito et al., 2005; You et al., 2013; Lovato et al., 2015) and in methane production (Yang et al., 2008; Hutñan et al., 2009; Viana et al., 2012; Lovato et al., 2012 and 2015; Baba et al., 2013; Silva et al., 2013; Zucoloto et al., 2018) by anaerobic treatment, which makes the Glycerin a substrate attractive due to easy biodegradation, anaerobic processes at appropriate pH and for being a great source of carbon (Sarma et al., 2013; Lovato et al., 2017).

Aiming at hydrogen production, Lovato et al., (2015) studied an AnSBBR with recirculation of the liquid phase (at 30°C with 3.5 L of working volume and treating 1.5 L per cycle) treating pure glycerin-based wastewater was applied to biohydrogen production. The applied volumetric organic load ranged from 7.7 to 17.1 $kgCOD.m^{-3}.d^{-1}$, combining different influent concentrations (3000, 4000 and 5000 $mgCOD.L^{-1}$) and cycle lengths (4 and 3 hours). The feed strategy used was to maintain the feeding time equal to half of the cycle time. The increase in the influent concentration and the decrease in cycle length improved the molar yield and molar productivity of hydrogen. The highest productivity (100.8 $molH_2.m^{-3}.d^{-1}$) and highest yield of hydrogen per load removed (20.0

molH$_2$.kgCOD^{-1}) were reached when the reactor operated with an AVOL of 17.1 kgCOD.m^{-3}.d^{-1}, with 68% of H$_2$ and only 3% of CH$_4$ in its biogas. It was also found that pretreatment of the sludge/inoculum does not influence the productivity/yield of the process and the use of crude industrial glycerin-based wastewater in relation to the pure glycerol-based wastewater substantially decreased the production and composition of the hydrogen produced.

In studies that sought methane production from glycerin treatment, Bezerra et al., (2011) studied the effect of applied organic volumetric load on the performance of an AnSBBRr operated in sequential batch with recirculation in the liquid phase. The authors reported organic matter removal efficiencies of 92%, 81%, 67% and 50% for applied organic loads of 1.5, 3.0, 4.5 and 6.0 gCOD.L^{-1}.d^{-1}, respectively. This decrease in removal efficiency was attributed to high concentration of total volatile acids in the effluent. In addition, the methane composition in the biogas was also impaired, ranging, respectively, from 72% to 64, 57 and 51%. However, the increase in organic load from 1.5 to 4.5 favored methane production from 29.5 to 55.5 NmL-CH4.gCOD-1. For the authors, the study suggested that for this reactor configuration the applied organic load limit should be less than 6.0 gCOD.L^{-1}.d^{-1}. As in the work of Vlassis et al., (2013), high concentrations of glycerol, as well as other impurities found in glycerin, may have an inhibitory effect on biogas production.

Comparing the work of Bezerra et al., (2011) with batch mode and of Lovato et al., (2012) with fed-batch mode, who worked with AnSBBR treating synthetic effluent from the biodiesel production process aiming at methane production, it is possible to notice a significant improvement with the use of fed-batch mode. Lovato et al., (2012) reached concentrations between 9.2 and 12.2 mmol.L^{-1} of methane in the biogas generated by the process against 1.4 to 2.4 mmol.L^{-1} found by Bezerra et al., (2011). Cheong and Hansen (2008) also found a significant improvement in the molar yield of methane using the fed-batch mode because while the maximum productivity reached in the batch mode was 32 mol CH$_4$.m^{-3}.d^{-1}, in the fed batch mode this value reached 141 molCH$_4$.m^{-3}.d^{-1}.

3.2. Thermophilic Treatment

In the case of thermophilic processes, it is expected that an increase in the reactor's capacity to operate at high organic loads consequently improves methane productivity. Using only glycerin as substrate, Zucoloto et al., (2018) identified a worsening performance of an AnSBBR operated at 55°C with mechanical agitation. The system obtained a decrease in organic matter removal efficiency from 77% to 55% when AVOL increased from 2.3 to 6.5 $gCOD.L^{-1}.d^{-1}$, probably due to the increase in the total volatile acids concentration in the effluent from 257 to 2229 $gHAc.L^{-1}$. On the other hand, biogas composition remained at approximately 65% of methane. In general, the best results obtained by the authors were at 4.3 $gCOD.L^{-1}.d^{-1}$ in which 66% organic matter removal efficiency and 51 $molCH_4.m^{-3}.d^{-1}$ were achieved. These results were similar to those obtained by Bezerra et al., (2011) and worse than those found in Lovato et al., (2012).

Therefore, analyzing these studies for the treatment of glycerin in AnSBBRs, mesophilic processes presented advantages when compared to thermophilic processes.

4. Treatment of Whey Based-Wastewater in Mesophilic and Thermophilic Conditions

Whey is the liquid remaining after the precipitation and removal of milk casein during the cheese-making process, and constitutes the largest by-product of the dairy industry (Gelegenis et al., 2007; Rico et al., 2015). It represents 90% of the milk volume used in the cheese production and retains 55% of milk nutrients. Untreated whey can be reused as animal feed, and if treated it presents a number of industrial applications such as in the manufacture of ice cream, cakes, dairy products, protein supplements, ethanol production, among other possibilities (Siso 1996). However, not all industries have the means and resources to reuse it, which is then discarded as waste (Bezerra et al., 2007).

Whey is composed mostly of water and usually contains 5-8% of solids. These solids are composed of 60-80% lactose, 10-20% protein and the rest is vitamins, fats, lactic acid and trace elements (Siso, 1996; Rico et al., 2015). Whey has a high organic load, BOD and COD values around, respectively, 27-60 and 50-102 g.L^{-1}, low pH (3.8-6.5), is highly biodegradable (close to 99%) and presents low alkalinity (around 2500 mg.L^{-1} CaCO$_3$) (Mawson, 1994; Ergüder et al., 2001; Prazeres et al., 2012). However, the composition of the whey may vary according to the animal origin and production process, so it depends of composition and quality of the milk used (Kavacik and Topaloglu, 2010).

According to the United States Department of Agriculture (FAS-USDA, 2017), the estimated production of cheese in Brazil for the year 2018 is 780 thousand tons. According to Carvalho et al., (2013), for every 1 L of bovine milk used to manufacture cheese, 0.873 ML of whey is generated, such that it is necessary to produce around 10 L of milk for each 1 kg of cheese. Applying this relation to the estimation of cheese production, it is expected a generation of 6.81 billion liters of whey. This volume presents a challenge for the dairy industry, since there are authors who suggest that approximately half of the whey produced in the world is not used. Even in cases where the whey is reused for protein recovery, the waste stream remaining still having a very high pollution load (mainly lactose contained in the permeate) and thus its direct disposal still constitutes a major environmental problem (Gelegenis et al., 2007).

Due to its composition and volume, the incorrect disposal of whey can cause an excess of oxygen consumption, impermeabilization, eutrophication, toxicity, etc. in the receiving environments. (Prazeres et al., 2012).

Anaerobic digestion is a good option to the whey treatment because of its high biodegradability. Besides that, the biological treatment of whey allows the production of biogas and possible energy utilization. However, due to its low alkalinity and easy degradation, high-load digestion processes can result in the accumulation of organic acids in the reactor, which can cause the process failure (Malaspina et al., 1996; Rico et al., 2015).

Because of the ease acid formation in anaerobic whey digestion, much of the work focused on biogas production is devoted to the hydrogen production. However, the delimitation of ideal conditions for the methanogenic microorganisms also allows the use of this substrate for the formation of methane. For this, some alternatives are used, such as whey dilution, alkalinity supplementation to the medium and / or the use of systems combined with separate acidogenic and methanogenic reactors (Mockaitis et al., 2006; Venetsaneas et al., 2009; Rico et al., 2015). Thus, the anaerobic digestion of whey is widely found in the literature with high values of removal efficiency of organic matter and conversion to methane (Yan et al., 1988, 1989; Malaspina et al., 1996; Ergüder et al., 2001; Mockaitis et al., 2006; Saddoud et al., 2007).

4.1. Mesophilic Treatment

The use of anaerobic reactors operated in sequential batch, with or without immobilized biomass, is easily found in literature. Ratusznei et al., (2003) demonstrated the feasibility of an ASBR with mechanical agitation treating whey at an organic load of 0.81-5.7 $gCOD.L^{-1}.d^{-1}$. The study demonstrated the need for low alkalinity supplementation (10% of bicarbonate mass per COD mass), but in the start-up of the reactor was needed more supplementation. The system was operated at 30 ° C, with cycle times of 8 hours and agitation of 200 rpm. It achieved high efficiency of removal organic matter, up to 96%.

A similar study was performed by Mockaitis et al., (2006) also using an ASBR with mechanical agitation, but with different configuration, cycle time of 8 hours and organic loads of 0.6-4.8 $gCOD.L^{-1}.d^{-1}$. Using up to 0.25 $gNaHCO_3.gCOD^{-1}$ of supplementation the study obtained COD removal efficiencies above 90% for all conditions, and reported the increase in methane production with the increase of organic load. It was noted a formation of a viscous substance similar to polymers in higher organic loads, as well as flotation of part of biomass, being this more intense in quantities of higher bicarbonate supplementation.

Damasceno et al., (2007) analyzed whey treatment in an AnSBBR with mechanical agitation at 30 ° C, cycle time of 8 hours, filling time of 10 minutes, 2 and 4 hours, applied volumetric organic loads of 2, 4, 8 and 12 $gCOD.L^{-1}.d^{-1}$. The organic matter removal efficiency was at least 81%.

Bezerra et al., (2009) also analyzed the treatment of whey in AnSBBR with recirculation of the liquid phase at 30°C, feeding time of 2, 4 and 6 hours, applied volumetric organic loads of 3, 6 and 12 $gCOD.L^{-1}.d^{-1}$. It was observed that in the condition that the feeding time was 2 hours the organic matter removal efficiency was high (about 98%). The authors also carried out a study of punctual overloads of 24 $gCOD.L^{-1}.d^{-1}$. After the application of this overloads, it was noted an increase of the organic acids, but it did not destabilized the system.

Lovato et al., (2016) studied the treatment of whey in an AnSBBR with recirculation of the liquid phase at 30°C and applied volumetric organic load of 7.5 $gCOD.L^{-1}.d^{-1}$. The reactor was operated in fed-batch, with cycle time of 8 hours and feeding time of 4 hours (50% cycle time). The system achieved 90% of soluble COD removal efficiency, methane productivity of 93.7 $molCH_4.m^{-3}.d^{-1}$, methane yield per applied COD of $molCH_4.kgCOD^{-1}$ and biogas was composed of 73% of methane.

Under similar conditions, but using an AnSBBR with mechanical agitation and applied volumetric organic load of 5.1 $gCOD.L^{-1}.d^{-1}$, Lovato et al., (2017) reported soluble COD removal efficiency of 94%, molar productivity of 72.9 $molCH_4.m^{-3}.d^{-1}$, methane yield per applied organic load of 14.4 $molCH_4.kgCOD^{-1}$ and 60% of methane in biogas.

Lovato et al., (2019) utilizou um AnSBBR com recirculação da fase líquida para tratar soro a 30°C. O reator foi operado em batelada com tempo de ciclo de 8 horas com uma carga de 6.7 $gCOD.L^{-1}.d^{-1}$. O sistema atingiu eficiência de remoção de COD de 87%, produtividade molar de metano de 76.3 $molCH_4.m^{-3}.d^{-1}$, rendimento de metano por massa aplicada de 11.3 $molCH_4.kgCOD^{-1}$ e composição de metano de 72%.

In this context, it is possible to note the viability of the application of AnSBBR in whey treatment and methane production under mesophilic conditions. Several studies have reported organic matter removals near to 90%, high methane productivities and yields. This works also reported stability and suitability of the systems when exposed to high organic loads, an interesting factor due to the high organic content of whey.

4.2. Thermophilic Treatment

Unlike mesophilic systems, studies that have explored the thermophilic digestion of whey for methane production are still scarce.

Fernández et al., (2015) studied the thermophilic anaerobic digestion of cheese whey using a single and a two stage configuration systems in a sequencing batch reactor (SBR), meaning hydrogen and methane production. The thermophilic SBR was used for digesting cheese whey in a single stage configuration and subsequently was used as a second stage digester for treating acid cheese whey. While the acidogenic reactor was operated at 35°C, the one stage system and the methanogenic reactor were operated at 55°C. The system was supplemented with sodium bicarbonate, monopotassium phosphate, potassium hydroxide and urea. Thermophilic anaerobic digestion of cheese whey was successfully achieved at an HRT of 8.3 days. The evaluation of the process under two-stage configuration (H_2 and CH_4) resulted in the need of higher HRT (12.5 days) for the methanogenesis phase. In the one-stage system, it was possible to achieve a removal organic matter efficiency of 87.4%, methane yield of 314 $mLCH_4.gCOD^{-1}$ and a biogas composition of 50-60% of methane. In the two-stage system, after 66 days of operation the methanogenic reactor presented drop in pH and methane production. This occurred in the condition that had the higher applied volumetric organic load (2.4 $gCOD.L^{-1}.d^{-1}$) and the reactor did not recover the performance even after a reduction of the applied organic volumetric load. The inhibition of methanogenesis was attributed to the non - adaptation of the methanogenic microorganisms to high concentrations of potassium and sodium, coming from the pH control solution of the system.

Siqueira et al., (2018) studied the thermophilic digestion of whey in an AnSBBR with mechanical agitation, operated at 55°C. The reactor was operated at applied organic loads of 6.2 to 30.3 $gCOD.L^{-1}.d^{-1}$, cycle time of 8 hours and sodium bicarbonate supplementation at ratios of 0.2-0.5 $gNaHCO_3.gCOD^{-1}$. The study reported that the methane productivity increased with the increase of applied volumetric organic load until the applied organic load achieve 24,7 $gCOD.L^{-1}.d^{-1}$. Applied organic loads higher than 24.7 $gCOD.L^{-1}.d^{-1}$ resulted in a decrease of methane yields and productivities. The best methane yield was achieved at applied organic load of 12,8 $gCOD.L^{-1}.d^{-1}$.

5. TREATMENT OF VINASSE BASED-WASTEWATER IN MESOPHILIC AND THERMOPHILIC CONDITIONS

In Brazil, the production of ethanol is mostly by sugar cane, being the largest producer of sugar cane in the world and the second in sugar and ethanol production. The production of ethanol doubled in the country in the last 20 years, reaching more than 106.6 million liters of produced ethanol in 2017 (ÚNICA, 2017).

The main residue of ethanol distillation is vinasse, which is generated in the proportion of 10 to 15 L per liter of ethanol produced. Vinasse is generated in the distillation column at a temperature of 85-90°C with a low pH (4.0 a 5.0), a dark brown color, high ash content and a high percentage of dissolved organic and inorganic matter (Wilkie et al., 2000; Pant and Adholeya, 2007; Vlissidis, 1993).

In addition, vinasse may also contain recalcitrant, antibacterial and toxic compounds, such as phenols, heavy metals and polymers, however vinasse presents a diverse characterization since the volume generated, its characteristics tend to vary according to the raw material, and the process productive used (Wilkie et al., 2000; Mohana et al., 2009; España-Gamboa et al., 2011).

According to Decloux and Bories (2002), sugarcane vinasse is composed of 2 to 10% solids, of which a great majority is composed of organic content. Among organic compounds, the literature suggests that these are mostly formed by organic acids, such as lactic and acetic acids, alcohols, such as ethanol and glycerol, and carbohydrates (Dowd et al., 1994; España-Gamboa et al., 2011).

The Works of Fuess et al., (2018) and Santos et al., (2014) reported high levels of iso-butyric, lactic, succinic and malic acids, followed by lower concentrations of ethanol, acetic, butyric and propionic acids. On the other hand, Janke et al., (2016) reports higher concentrations of acetic acid, lactic acid and ethanol. Both works still denote high concentrations of nutrients such as nitrogen, phosphorus, potassium, sulfate, calcium and magnesium.

Several forms of vinasse disposal have been studied over recent years, including biological treatment, fertirrigation, anaerobic digestion for methane production, yeast production and even animal feed utilization (Robertiello, 1982). Since the end of 1970, fertirrigation has been the most widely used method of disposal of vinasse in Brazil. This is due mainly to its nutritional composition and because it is an economical, efficient, fast-applied and dispensable alternative to complex technologies (Santana et al., 2008; Moraes et al., 2015).

However, this long-term practice, without dosage criteria and planning of irrigation systems, can promote changes in soil physical properties and contamination of local water resources. Impacts include: (i) soil salinization; (ii) organic overload; (iii) excessive fertilization; (iv) soil acidification; (v) interference in the photosynthesis process of the water bodies affected with the leaching; (vi) inhibition of seed germination; and (vii) increased CO_2 and N_2O emissions in areas of application with sugar cane cultivation (Silva et al., 2007; Oliveira et al., 2013; Fuess and Garcia, 2014).

Anaerobic digestion is an attractive alternative for the treatment of vinasse, since it allows the reduction of polluting potential, minimizing the impacts caused by the fertirrigation, to recover the biogas generated as final product of the process. Even after the anaerobic digestion process, part of the nutrients present in the vinasse (N, P, K, Fe, Zn, Mn, Cu, Mg) are still in the effluent, thus maintaining their potential for fertirrigation (Wilkie et al., 2000; Christofoletti et al., 2013; Fuess and Garcia, 2014).

5.1. Mesophilic Treatment

The use of anaerobic reactors operated in sequential batch and/or fed-batch for vinasse treatment has been studied over the last years. Albanez et al., (2016a), for example, evaluated the anaerobic digestion of sugarcane vinasse in AnSBBR with mechanical agitation at 30°C and cycle length of 8 hours. The work evaluated the influence of the increase in applied volumetric organic load (1.18 to 5.54 $gCOD.L^{-1}.d^{-1}$) in the reactor performance, in relation to organic matter removal and methane yield and productivity. The best results were achieved in the condition with the higher applied organic load (organic matter removal efficiency of 83%, methane content in the biogas of 77%, methane productivity of 43.4 $molCH_4.m^{-3}.d^{-1}$ and yield of 9.47 $molCH_4.kgCOD^{-1}$).

Almeida et al., (2017) analyzed the anaerobic digestion of vinasse in AnSBBR reactor with mechanical agitation at 30°C and cycle time of 8 hours. The work evaluated the influence of the increase of the organic load (1.09 a 10.1 $gCOD.L^{-1}.d^{-1}$), the feeding strategy (batch and fed batch) and the temperature (30 and 45°C) in the reactor performance indicators. The increase in the applied volumetric organic load increased methane productivity (123 $molCH_4.m^{-3}.d^{-1}$). However, the higher methane yield (13,5 $mmolCH_4.gCOD^{-1}$) and the best biogas composition (79% of methane) were achieved with a applied organic load of 8,18 $gCOD.L^{-1}.d^{-1}$. It was noted that the operation in batch mode was better to reactor performance in relation to organic matter removal and methane productivity and yield.

Volpini et al., (2018) studied the application of an AnSBBR with recirculation of liquid phase operated in sequencing batch and fed-batch, in the treatment of wastewater from biohydrogen production process from vinasse (sugar cane stillage) in mesophilic condition (30°C) and cycle length of 8 hours. The acidogenic reactor that produced the influent for the reactor studied by Volpini et al., (2018) was presented in work of Albanez et al., (2016b). Volpini et al., (2018) studied the influence of the increase of applied volumetric organic load (1.9 a 8.8 $gCOD.L^{-1}.d^{-1}$) in reactor performance indicators. It achieved organic matter removal efficiency between 71 e 89% and high methane productivity and yield, the highest methane productivity achieved was 133 $molCH_4.m^{-3}.d^{-1}$.

5.2. Termophilic Treatment

Vinasse is generated in the distillation column at a temperature of 85-90°C. When vinasse treatment is applied in a mesophilic condition, it is necessary to precool this effluent, so it becomes attractive to apply the treatment in a thermophilic condition, since in addition to dispensing with the cooling stage, it increases the productivity of methane and the yield factor between generated methane and substrate consumed (Vlissidis, 1992; Song et al., 2004; Tatara et al., 2005; Ueno et al., 2007; Yang et al., 2008; Ferraz Júnior et al., 2016).

In thermophilic processes high temperatures accelerate chemical and biological reaction rates, but the optimal growth temperature of the microorganisms responsible for the biotransformations must be respected (Van Lier, 1995; Ribas et al., 2009). There are few treatment plants that use anaerobic reactors in thermophilic condition to treat vinasse, but studies show that in the thermophilic temperature both the degradation and the stabilization of the organic matter is faster in relation to the mesophilic sludge (Wohlt et al., 1990).

Souza et al., (1992) operated a semi-industrial scale UASB in thermophilic condition, achieving an organic matter removal efficiency of 72% with AVOL of 30 $kgCOD.m^{-3}.d^{-1}$. Harada et al., (1993) studied the behavior of anaerobic sludge at 55°C and 65°C, it was observed the stability of the system (34 $kgCOD.L^{-3}.d^{-1}$ and 16 $kgCOD.m^{-3}.d^{-1}$).

Wilkie et al., (2000) studied the performance of UASB reactors in mesophilic and thermophilic conditions by treating vinasse and concluded that the thermophilic reactor achieves better results in the removal of biodegradable organic matter when was applied high organic load.

Luo et al., (2010) studied anaerobic hydrogen and methane production from cassava stillage in continuously stirred tank reactor (CSTR) in mesophilic and thermophilic conditions. Results showed that under high organic loading rate (OLR) (>10 $gVS.L^{-1}d^{-1}$), the two-phase thermophilic CSTR for hydrogen and methane production was stable with hydrogen and methane yields of 56.6 $mLH_2.gVS^{-1}$ and 249 $mLCH_4.gVS^{-1}$. The one-phase thermophilic CSTR for methane production failed due to the accumulation of both acetate and propionate, leading to the pH lower than 6. Instead of propionate alone, the accumulations of both acetate and propionate were found to be related to the breakdown of methane reactor.

Ferraz Júnior et al., (2016) evaluated the performance of thermophilic methanogenic up-flow sludge blanket reactors (UASB) operating continuously in single-stage (UASB I) and two-stage (UASB II) systems vinasse in relation to organic matter removal and energy production. The maximum methane yields were 250.9 NmL-CH_4. $CH_4.gCOD_{removed}^{-1}$ and 316.0 NmL-$CH_4.gCOD_{removed}^{-1}$ for UASB I and II, respectively, corresponding to 71.7% and 90.3% of the maximum theoretical methane yield (350 NmL-$CH_4.gCOD_{removed}^{-1}$). The organic matter removal efficiency was about 70% for the two systems.

The studies using AnSBBR reactor for thermophilic treatment of vinasse are scarce in the literature. Almeida et al., (2017) assessed the feasibility of an AnSBBR with mechanical stirring for vinasse treatment and biomethane production. The cycle time was 8 h, with feeding times of 10 min (batch mode) and 240 min (fed-batch mode), the temperatures studied were 30ºC and 45ºC. The maximum molar productivity and yield of methane were 123.4 $molCH4.m^{-3}.d^{-1}$ and 13.8 $mmolCH4.gCOD^{-1}$ (88% of the theoretical), respectively. These parameters were lower at 45°C (35.0 $molCH_4.m^{-3}.d^{-1}$ and 7.10 $mmolCH_4.gCOD^{-1}$) than at 30°C and the organic matter removal efficiency was about 46%. The worsening in reactor performance may be related to both the temperature increase strategy adopted and the value of 45°C being the value corresponding to the transition range from the mesophilic to the thermophilic condition.

On the other hand, Albuquerque et al., (2019) operated an AnSBBR reactor at 55°C and reported that increasing the organic load (5 - 25 $gCOD.L^{-1}.d^{-1}$) and changing the feeding strategy form batch to fed-batch mode improved reactor performance, organic matter removal efficiency was 79% and the methane yield was 14.8 $mmolCH_4.gCOD^{-1}$, close to the theoretical value (15.6 $mmolCH_4.gCOD^{-1}$). The authors did a comparative analysis with studies that treated vinasse in AnSBBR in mesophilic and thermophilic conditions, they verified that the thermophilic reactor obtained, generally, better results, even when operated with lower organic loads.

CONCLUSION

The presented studies confirm the operational flexibility of the process, in which satisfactory results were obtained both in batch and fed-batch, where improvement in system performance was achieved using feed time equal to 50% of the cycle length (generally 8h).

Regarding applied organic loading rate, studies revealed the possibility of application of high loads in the treatment of the agroindustrial residues glycerin, whey and vinasse. In the case of glycerin, organic loading rates of 1.5 to 7.5 $gCOD.L^{-1}.d^{-1}$ were reported with COD removal efficiencies ranging from 92 to 50%, and yields of up to 21.4 $molCH_4.m^{-3}.d^{-1}$ for systems and densities of 3.3 to 6.5 $gCOD.L^{-1}.d^{-1}$ with efficiencies of 77 to 55% and yields of up to 51 $molCH_4.m^{-3}.d^{-1}$ for thermophilic systems.

For whey based wastewaters at organic loading rates of 0.8 to 12 $gCOD.L^{-1}.d^{-1}$, organic matter removal efficiencies were reported to be 98-81% with methane yields up to 93 $molCH_4.m^{-3}.d^{-1}$ for mesophilic systems, and, for thermophilic systems, at organic loading rates of 1.2 to 24.7 $gCOD.L^{-1}.d^{-1}$, organic matter removal efficiencies greater than 85% were achieved with methane yields up to 324 $molCH_4.m^{-3}.d^{-1}$.

Regarding vinasse based wastewaters, authors reported organic loading rates ranging from 1 to 36 $gCOD.L^{-1}.d^{-1}$ with removal efficiencies of 98 to 71% and yields up to 123 $molCH_4.m^{-3}.d^{-1}$ for mesophilic systems, and organic loading rates up to 25 gCOD.L^{-1}.d^{-1} with efficiencies up to 79% and yields up to 352 molCH$_4$.m^{-3}.d^{-1}. Therefore, of the three wastewaters, glycerin presented the greater difficulty at high organic loading rates and better results for mesophilic ranges of operation. On the other hand, studies with whey presented better performances at higher organic loads when in thermophilic systems, indicating even higher productivities at higher temperatures. Vinasse, on the other hand, at high organic loads achieved high values of organic matter removal efficiency in both temperature ranges, but higher methane productivities were reporter in thermophilic systems.

The analysis showed a high performance and stability by AnSBBRs that aimed at organic matter degradation and methane production by digesting the three wastewaters highlighted in this work: glycerin, whey and vinasse.

Despite the promising results found in literature regarding methane production in AnSBBRs both in mesophilic and thermophilic conditions, the reports of the application of this kind of reactor in pilot or full scale for the treatment of agroindustrial wastes are still scarce.

ACKNOWLEDGMENTS

This study was supported by the São Paulo Research Foundation (FAPESP: #2015/06246-7), the National Council for Scientific and Technological Development (CNPq: #443181/2016-0), and the Coordination for the Improvement of Higher Education Personnel (CAPES).

REFERENCES

Albanez, R., Chiaranda, B. C., Ferreira, R. G., França, A. L. P., Honório, C. D., Rodrigues, J. A. D., Ratuszei, S. M., Zaiat, M. Anaerobic

Biological Treatment of Vinasse for Environmental Compliance and Methane Production. *Applied Biochemistry and Biotechnology*, 178, 21-43, 2016a.

Albanez, R., Lovato, G., Zaiat, M., Ratusznei, S. M., Rodrigues, J. A. D. Optimization, metabolic pathways modeling and scale-up estimative of an AnSBBR applied to biohydrogen production by co-digestion of vinasse and molasses. *International Journal of Hydrogen Energy*, 41, 20473-20484, 2016b.

Albuquerque, J. N., Orellana, M. R., Ratusznei, S. M., Rodrigues, J. A. D. Thermophilic biomethane production by vinasse in na AnSBBR: start-up strategy and performance optimization. *Brazilian Journal of Chemical Engineering*, 2019. (in press)

Almeida, W. A., Ratusznei, S. M., Zaiat, M., Rodrigues, J. A. D. AnSBBR applied to biomethane production for vinasse treatment: effects of organic loading, feed strategy and temperature. *Brazilian Journal of Chemical Engineering*, 34, 759-773, 2017.

Agência Nacional Do Petróleo, Gás Natural E Biocombustíveis (2017). *Anuário estatístico brasileiro do petróleo, gás natural e biocombustíveis*. Available from: <http://www.anp.gov.br>. Last access: November 6, 2018. [*Brazilian Statistical Yearbook of Petroleum, Natural Gas and Biofuels*]

Archilha, N. C., Canto, C. S., Ratusznei, S. M., Rodrigues, J. A. D., Zaiat, M., Foresti, E. Effect of feeding strategy and COD/sulfate ratio on the removal of sulfate in an AnSBBR with recirculation of the liquid phase. *Journal of Environmental Management*, 91, 1756-1765, 2010.

Baba, Y., Tada, C., Watanabe, R., Fukuda, Y., Chida, N., Nakai, Y. Anaerobic digestion of crude glycerol from biodiesel manufacturing using a large-scale pilot plant: Methane production and application of digested sludge as fertilizer. *Bioresource Technology*, 140, 342-348, 2013.

Bagley, D. M, Brodkorb, T. S. Modeling microbial kinetics in an anaerobic sequencing batch reactor: model development and experimental validation. *Water Environment Research*, 71, 1320–1332, 1999.

Bezerra, R. A., Rodrigues, J. A. D., Ratusznei, S. M., Zaiat, M., Foresti, E. Effects of feed time, organic loading and shock loads in anaerobic whey treatment by an AnSBBR with circulation. *Applied Biochemistry and Biotechnology*, 157, 140-158, 2009.

Bezerra, R. A., Rodrigues, J. A. D., Ratusznei, S. M., Canto, C. S. A., Zaiat, M. Effect of organic load on the performance and methane production of an AnSBBR treating effluent from biodiesel production. *Applied Biochemistry and Biotechnology*, 165, 347-368, 2011.

Bezerra, R. A., Rodrigues, J. A. D., Ratusznei, S. M., Zaiat, M., Foresti, E. Whey treatment by AnSBBR with circulation: effects of organic loading, shock loads, and alkalinity supplementation. *Applied Biochemistry and Biotechnology*, 143, 257-275, 2007.

Borges, A. C., Siman, R. R., Rodrigues, J. A. D., Ratusznei, S. M., Zaiat, M., Foresti, E., Borzani, W. Stirred anaerobic sequencing batch reactor containing immobilized biomass: a behavior study when submitted to different fill times. *Water Science and Technology*, 49, 311-318, 2004.

Bravo, I. S. M., Lovato, G., Rodrigues, J. A. D., Ratusznei, S. M., Zaiat, M. Biohydrogen production in an AnSBBR treating glycerin-based wastewater: effects of organic loading, influent concentration and cycle time. *Applied Biochemistry and Biotechnology*, 175, 1892-1914, 2015.

Cakir, F. Y, Stenstrom, M. K. C Greenhouse gas production: A comparison between aerobic and anaerobic wastewater treatment technology. *Water Research*, 4197-4203, 2005.

Camargo, E. F. M., Ratusznei, S. M., Rodrigues, J. A. D., Zaiat, M., Borzani, W. Treatment of low-strength wastewater using immobilized biomass in a sequencing batch external loop reactor: influence of the medium superficial velocity on the stability and performance. *Brazilian Journal of Chemical Engineering*, 19, 267-275, 2002.

Carvalho, F., Prazeres, A. R., Rivas, J. Cheese whey wastewater: Characterization and treatment. *Science of The Total Environment*, 385-396, 2013.

Christofoletti, C. A., Escher, J. P., Correia, J. E., Marinho, J. F. U., Fontanetti, C. S. Sugarcane vinasse: Environmental implications of its use. *Waste Management*, 33, 2752-2761, 2013.

Cubas, S. A., Foresti, E., Rodrigues, J. A. D., Ratusznei, S. M., Zaiat, M. Effect of impeller type and stirring frequency on the behavior of na AnSBBR in the treatment of low-strength wastewater. *Bioresource Technology*, 102, 889-893, 2011.

Damasceno, L. H. S., Rodrigues, J. A. D., Ratusznei, S. M., Zaiat, M., Foresti, E. Effect of mixing mode on the behavior of an ASBBR with immobilized biomass in the treatment of cheese whey. *Brazilian Journal of Chemical Engineering*, 25, 291-298, 2008.

Damasceno, L. H. S., Rodrigues, J. A. D., Ratusznei, S. M., Zaiat, M., Foresti, E. Effects of feeding time and organic loading in an anaerobic sequencing batch biofilm reactor (ASBBR) treating diluted whey. *Journal of Environmental Management*, 85, 927-935, 2007.

Decloux, M, Bories, A. Stillage treatment in the French alcohol fermentation industry. *International Sugar Journal*, 104, 509-517, 2002.

Döll, M. M. R., Foresti, E. Efeito do bicarbonato de sódio no tratamento de vinhaça em AnSBBR operado a 55 e 35°C. *Engenharia Sanitaria e Ambiental*, 15, 275-282, 2010. [Effect of sodium bicarbonate on vinasse treatment in AnSBBR operated at 55 and 35oC. *Sanitary and Environmental Engineering*]

Dowd, M. K., Johansen, S. L., Cantarella, L., Reilly, P. J. Low molecular weight organic composition of ethanol stillage from sugarcane molasses, citrus waste, and sweet whey. *Journal of agricultural and food chemistry*, 42, 283-288, 1994.

Ergüder, T. H., Tezel, U., Güven, E., Demirer, G. N. Anaerobic biotransformation and methane generation potential of cheese whey in batch and UASB reactors. *Waste Management*, 21, 643-650, 2001.

España-Gamboa, E., Mijangos-Cortes, J., Barahona-Perez, L., Dominguez-Maldonado, J., Hernández-Zárate, G., Alzate-Gaviria, L. Vinasses: characterization and treatments. *Waste Management Research*, 29, 1235-1250, 2011.

FAS-USDA (2017) *Dairy, Milk, Fluid, Dairy, Cheese, Dairy, Butter, Dairy, Dry Whole Milk Powder Annual Dairy Report* |Brasilia|Brazil|11/29/2017. Available from: <https://gain.fas.usda.gov/ RecentGAINPublications/DairyandProductsAnnual_Brasilia_Brazil_1 0-26-2017.pdf>. Last access July 12, 2018.

Fernández, C., Cuetos, M. J., Martínez, E. J., Gómez, X. Thermophilic anaerobic digestion of cheese whey: Coupling H2 and CH4 production. *Biomass and Bioenergy*, 81, 55-62, 2015.

Ferraz Júnior, A. D. N., Koyama, M. H., Araújo Júnior, M. M., Zaiat, M. Thermophilic anaerobic digestion of raw sugarcane vinasse. *Renewable Energy*, 89, 245-252, 2016.

Fuess, L. T., Garcia, M. L. Implications of stillage land disposal: A critical review on the impacts of fertigation. *Journal of Environmental Management*, 145, 210-229, 2014.

Fuess, L. T., Garcia, M. L., Zaiat, M. Seasonal characterization of sugarcane vinasse: Assessing environmental impacts from fertirrigation and the bioenergy recovery potential through bio-digestion. *Science of The Total Environment*, 634, 29-40, 2018.

Gelegenis, J., Georgakakis, D., Angelidaki, I., Mavris, V. Optimization of biogas production by co-digesting whey with diluted poultry manure. *Renewable Energy*, 32, 2147-2160, 2007.

Harada, H., Uemura, S., Chen, A., Jayadevan, J. Anaerobic treatment of a recalcitrant distillery wastewater by a thermophilic UASB reactor. *Bioresource Technology*, 55, 215-221, 1996.

Hirl, P. J., Irvine, R. L. Reductive dechlorination of perchloroethylene (PCE) using anaerobic sequencing batch biofilm reactors (AnSBBR). In: *51ˢᵗ Purdue Industrial Waste Conference Proceesings*. Ann Harbor Press, Chelsea, MI, Purdue Industrial Waste, 289-295, 1996.

Hutñan, M., Kolesárová, N., Bodík, I., Czolderová, M. Long-term mono-digestion of crude glycerol in a UASB reactor. *Bioresource Technology*, 130, 88-96, 2013.

Ito, T., Nakashimada, Y., Senba, K., Matsui, T., Nichio, N. Hydrogen and ethanol production from glycerol-containing wastes discharged after biodiesel manufacturing process. *Journal of Bioscience and Bioengineering.* 100, 260-265, 2005.

Janke, L., Leite, A. F., Batista, K., Silva, W., Nikolausz, M., Nelles, M., Stinner, W. Enhancing biogas production from vinasse in sugarcane biorefineries: Effects of urea and trace elements supplementation on process performance and stability. *Bioresource Technology*, 217, 10-20, 2016.

Johnson, D. T., Taconi, K. A. The glycerin glut: options for the value-added conversion of crude glycerol resulting from biodiesel production. *Environmental Progress*, 26, 388-348, 2007.

Kavacik, B., Topaloglu, B. Biogas production from co-digestion of a mixture of cheese whey and dairy manure. *Biomass and Bioenergy*, 34, 1321-1329, 2010.

Larsen, A. C. (2009). *Codigestão anaeróbia de glicerina bruta e efluente de fecularia.* Available from: <http://www.dominiopublico.gov.br/download/texto/cp114920.pdf>. Last access: November 3, 2018. *[Anaerobic codigestion of crude glycerin and starch effluent]*

Lima, D. M. F., Lazaro, C. Z., Rodrigues, J. A. D., Ratusznei, S. M., Zaiat, M. Optimization performance of an AnSBBR applied to biohydrogen production treating whey. *Journal of Environmental Management*, 169, 191-201, 2016.

Lo, Y., Chen, X., Huang, C., Yuan, Y., Chang, J. Dark fermentative hydrogen production with crude glycerol from biodiesel industry using indigenous hydrogen-producing bacteria. *International Journal of Hydrogen Energy*, 38, 15815-15822, 2013.

Lovato, G., Albanez, R., Albuquerque, J. N., Cola, P., Celestino, R. S., Vogel, S. E., Fukuyama, M., Hirata, F. E., Saito, F. H., Ratusznei, S., Rodrigues, J. A. D. Novel insights into the co-digestion of whey with glycerin in an AnSBBR: Influent composition and concentration, cycle length and feed strategy effect. *Advances in Environmental Research*, 161-182, 2017.

Lovato, G., Albanez, R., Lima, D. M. F., Bravo, I. S. M., Almeida, W. A., Ratuznei, S. M., Rodrigues, J. A. D. Application and environmental compliance of anaerobic sequencing batch reactors applied to hydrogen/methane bioenergy production. *Watewater Treatment, Management Strategies and Environmental/Health Impacts*, p. 111-161, 2015.

Lovato, G., Albanez, R., Triveloni, M., Ratusznei, S. M., Rodrigues, J. A. D. Methane production by co-digesting vinasse and whey in an AnSBBR: Effect of mixture ratio and feed strategy. *Applied Biochemistry and Biotechnology*, 187, 28-46, 2019.

Lovato, G., Bezerra, R. A., Rodrigues, J. A. D., Ratusznei, S. M., Zaiat, M. Effect of feed strategy on methane production and performance of an AnSBBR treating effluent from biodiesel production. *Applied Biochemistry and Biotechnology*, 166, 2007-2029, 2012.

Lovato, G., Ratusznei, S. M., Rodrigues, J. A. D., and Zaiat, M. Co-digestion of Whey with Glycerin in an AnSBBR for Biomethane Production. *Applied Biochemistry and Biotechnology*, 178, 126-143, 2016.

Luo, G., Xie, L., Zou, Z., Zhou, Q., Wang, J. Fermentative hydrogen production from cassava stillage by mixed anaerobic microflora: effects of temperature and pH. *Applied Energy*, 87, 3710-3717, 2010.

Malaspina, F., Cellamare, C. M., Stante, L., and Tilche, A. Anaerobic treatment of cheese whey with a downflow-upflow hybrid reactor. *Bioresource Technology*, 55, 131-139, 1996.

Manssouri, M., Rodrigues, J. A. D., Ratusznei, S. M., Zaiat, M. Effects of Organic Loading, Influent Concentration, and Feed Time on Biohydrogen Production in a Mechanically Stirred AnSBBR Treating Sucrose-Based Wastewater. *Applied Biochemistry and Biotechnology*, 171, 1832–1854, 2013.

Mawson, A. J. Bioconversions for whey utilization and waste abatement. *Bioresource Technology*, 47(3), 195–203, 1994.

Mendes, D. B., Serra, J. C. V. Glicerina: uma abordagem sobre a produção e o tratamento. *Revista Liberato*, 13, 2012. [Glycerin: an approach to production and treatment. *Liberato Magazine*]

Mockaitis, G., Ratusznei, S. M., Rodrigues, J. A. D., Zaiat, M., Foresti, E. Anaerobic whey treatment by a stirred sequencing batch reactor

(ASBR): effects of organic loading and supplemented alkalinity. *Journal of Environmental Management*, 79, 198-206, 2006.

Mohan, S. V., Babu, V. L., Bhaskar, Y. V., Sarma, P. N. Influence of recirculation on the performance of anaerobic sequencing batch biofilm reactor (AnSBBR) treating hypersaline composite chemical wastewater. *Bioresource Technology,* 98, 1373-1379, 2007.

Mohana, S., Acharya, B. K., Madamwar, D. Distillery spent wash: Treatment technologies and potential applications. *Journal of Hazardous Materials*, 163, 12-25, 2009.

Moraes, B. S., Zaiat, M., Bonomi, A. Anaerobic digestion of vinasse from sugarcane ethanol production in Brazil: Challenges and perspectives. *Renewable and Sustainable Energy Reviews,* 44, 888-903, 2015.

Novaes, L. F., Borges, L. O., Rodrigues, J. A. D., Ratusznei, S. M., Zaiat, M., Foresti, E. Effect of Fill Time on the Performance of Pilot-scale ASBR and AnSBBR Applied to Sanitary Wastewater Treatment. Applied *Biochemistry and Biotechnology*, 162, 885–899, 2010.

Oliveira, B. G., Carvalho, J. L. N., Cerri, C. E. P., Cerri, C. C., Feigl, B. J. Soil greenhouse gas fluxes from vinasse application in Brazilian sugarcane areas. *Geoderma*, 200, 77-84, 2013.

Oliveira, D. S., Prinholato, A. C., Ratusznei, S. M., Rodrigues, J. A. D., Zaiat, M., Foresti, E. AnSBBR applied to the treatment of wastewater from a personal care industry: effect of organic load and fill time. *Journal of Environmental Management*, 90, 3070-3081, 2009.

Pant, D., Adholeya, A. Biological approaches for treatment of distillery wastewater: A review. *Bioresource Technology*, 98, 2321-2334, 2007.

Prazeres, A. R., Carvalho, F., Rivas, J. Cheese whey management: A review. *Journal of Environmental Management*, 110(Supplement C), 48–68, 2012.

Ramos, A. C. T., Ratusznei, S. M., Rodrigues, J. A. D., Zaiat, M. Mass transfer improvement of a fixed-bed anaerobic sequencing batch reactor with liquid-phase circulation. *Interciencia*, 28, 214-219, 2003.

Ratusznei, S. M., Rodrigues, J. A. D., Camargo, E. F. M., Zaiat, M., Borzani, W. Feasibility of a stirred anaerobic sequencing batch reactor containing immobilized biomass for wastewater treatment. *Bioresource Technology*, 75(2), 127–132, 2000.

Ratusznei, S. M., Rodrigues, J. A. D., Zaiat, M. Operating feasibility of anaerobic whey treatment in a stirred sequencing batch reactor containing immobilized biomass. *Water Science and Technology: A Journal of the International Association on Water Pollution Research*, 48, 179-186, 2003.

Ribas, M. M. F., Chinalia, F. A., Pozzi, E., Foresti, E. Microbial succession within an anaerobic sequencing batch biofilm reactor (ASBBR) treating cane vinasse at 55°C. *Brazilian Archives of Biology and Technology*, 52, 1027-1036, 2009.

Rico, C., Muñoz, N., Fernández, J., Rico, J. L. High-load anaerobic co-digestion of cheese whey and liquid fraction of dairy manure in a one-stage UASB process: Limits in co-substrates ratio and organic loading rate. *Chemical Engineering Journal*, 262, 794-802, 2015.

Rivero, M., Solera, R., Perez, M. Anaerobic mesophilic co-digestion of sewage sludge with glycerol: Enhanced biohydrogen production. *International Journal Hydrogen Energy*, 39, 2481-2488, 2014.

Robertiello, A. Upgrading of agricultural and agro-industrial wastes: The treatment of distillery effluents (vinasses) in Italy. *Agricultural Wastes*, 4, 387-395, 1982.

Rodrigues, J. A. D., Oliveira, R. P., Ratusznei, S. M., Zaiat, M., Foresti, E. AnSBBR Applied to a personal care industry wastewater treatment: effects of fill time, volume treated per cycle, and organic load. *Applied Biochemistry and Biotechnology*, 163, 127-142, 2011.

Saddoud, A., Hassaïri, I., Sayadi, S. Anaerobic membrane reactor with phase separation for the treatment of cheese whey. *Bioresource Technology*, 98, 2102–2108, 2007.

Santana, V. S., Machado, N. R. C. F. Photocatalytic degradation of the vinasse under solar radiation. *Catalysis Today*, 133, 606-610, 2008.

Santos, S. C., Rosa, P. R. F., Sakamoto, I. K., Varesche, M. B. A., Silva, E. L. Organic loading rate impact on biohydrogen production and microbial communities at anaerobic fluidized thermophilic bed reactors treating sugarcane stillage. *Bioresource Technology*, 159, 55–63, 2014.

Sarma, S. J., Brar, S. K., Sydney, E. B., Bihan, Y. L., Buelna, G., Soccol, C. R. Microbial hydrogen production by bioconversion of crude glycerol: a review. *International Journal Hydrogen Energy*, 37, 6473-6490, 2012.

Silva, M. A. S., Griebeler, N. P., Borges, L. C. Uso de vinhaça e impactos nas propriedades do solo e lençol freático. *Revista Brasileira de Engenharia Agrícola e Ambiental*, 11, 108-114, 2007. [Use of vinasse and impacts on soil and ground water properties. *Brazilian Journal of Agricultural and Environmental Engineering*]

Silva, R. C., Ratusznei, S. M., Rodrigues, J. A. D., Zaiat, M. Anaerobic treatment of industrial biodiesel wastewater by an ASBR for methane production. *Applied Biochemistry and Biotechnology*, 170, 105-118, 2013.

Siqueira, T. S., Albuquerque, J. N., Ratusznei, S. M., Rodrigues, J. A. D. Biomethane production from whey treatment in an ansbbr at thermophilic condition. In: Joanne Castillo (ed.), *Bioenergy: Prospects, Applications and Future Directions*. Hauppauge, NY, Nova Science Publishers, p. 13-41, 2018.

Siso, M. I. G. The biotechnological utilization of cheese whey: A review. *Bioresource Technology*, 57, 1-11, 1996.

Song, Y., Kwon, S., Woo, J. Mesophilic and thermophilic temperature co-phase anaerobic digestion compared with single-stage mesophilic and thermophilic digestion of sewage sludge. *Water Research*, 38, 1653-1662, 2004.

Souza, M. E., Fuzaro, G., Polegato, A. R. Thermophilic anaerobic digestion of vinasse in pilot plant UASB reactor. *Water Science & Technology*, 25, 213-222, 1992.

Sung, S, Dague, R. R. Laboratory Studies on the anaerobic sequencing batch reactor. *Water Environment Research*, 67, 294-301, 1995.

Tatara, M., Yamazawa, A., Ueno Y., Fukui, H., Goto, M., Sode, K. High-rate thermophilic methane fermentation on short-chain fatty acids in a down-flow anaerobic packed-bed reactor. *Bioprocess and Biosystem Engineering*, 27, 105-113, 2005.

Ueno, Y., Fukui, H., Goto, M. Operation of a two-stage fermentation process producing hydrogen and methane from organic waste. *Environmental Science Technology*, 41, 1413-1419, 2007.

UNICA, União da Indústria de Cana-de-açúcar. *Safra da cana de açúcar, produção de etanol, consumo de etanol.* Available from: <www.unica.com.br>. Last access: April 6, 2018. [Union of the Sugarcane Industry. *Sugarcane harvest, ethanol production, ethanol consumption*]

Van Lier, J.B. *Thermophilic anaerobic wastewater treatment: temperature aspects and process stability.* PhD Thesis, Wageningen University, 181 p, 1995.

Venetsaneas, N., Antonopoulou, G., Stamatelatou, K., Kornaros, M., Lyberatos, G. Using cheese whey for hydrogen and methane generation in a two-stage continuous process with alternative pH controlling approaches. *Bioresource Technology,* 100, 3713-3717, 2009.

Viana, M. B., Freitas, A. V., Leitão, R. C., Pinto, G. A. S., Santaella, S. T. Anaerobic digestion of crude glyceron: A review. *Environment Technology*, 1, 81-92, 2012.

Vlassis, T., Stamatelatou, K., Antonopoulou, G., Lyberatos, G. Methane production via anaerobic digestion of glycerion: a comparasion of conventional (CSTR) and high-rate (PABR) digesters. *Journal of Chemical Technology and Biotechnology*, 88, 2000-2006, 2013.

Vlissidis, A., Zouboulis, A. Thermophilic anaerobic digestion of alcohol distillery wastewaters. *Bioresource Technology*, 43, 131-140, 1993.

Volpini, V., Lovato, G., Albanez, R., Ratusznei, S. M., Rodrigues, J. A. D. Biomethane generation in an AnSBBR treating effluent from the biohydrogen production from vinasse: Optimization, metabolic pathways modeling and scale-up estimation. *Renewable Energy*, 116, 288-298, 2018.

Wilkie, A. C., Riedesel, K. J., Owens, J. M. Stillage characterization and anaerobic treatment of ethanol stillage from conventional and cellulosic feedstocks. *Biomass and Bioenergy*, 19, 63-102, 2000.

Wohlt, J. E., Frobish, R. A., Davis, C. L., Bryant, M. P., Mackie, R. Thermophilic methane production from dairy cattle waste. *Biological Wastes*, 32, 193-207, 1990.

Yan, J. Q., Liao, P. H., Lo, K. V. Methane production from cheese whey. *Biomass*, 17, 185-202, 1988.

Yan, J. Q., Lo, K. V, Liao, P. H. Anaerobic digestion of cheese whey using up-flow anaerobic sludge blanket reactor. *Biological Wastes*, 27, 289-305, 1989.

Yang, Y., Tsukahara, K., Sawayama, S. Biodegradation and methane production from glycerol-containing synthetic wastes with fixed-bed bioreactor under mesophilic and thermophilic anaerobic conditions. *Process Biochemistry*, 43, 362-367, 2008.

Zaiat, M., Rodrigues, J. A. D., Ratusznei, S. M., Camargo, E. F. M., Borzani, W. Anaerobic sequencing batch reactors for wastewater treatment: a developing technology. *Applied Microbiology and Biotechnology*, 55, 29-35, 2001.

Zucoloto, N. F., Lovato, G., Albanez, R., Ratusznei, S. M., Rodrigues, J. A. D. *Thermophilic biomethane production by co-digesting glycerin and molasses in an AnSBBR: Effects of composition and applied organic load.* Nova Scrience, p. 215-248, 2019.

INDEX

Related Nova Publications

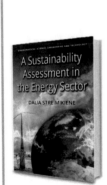